# HONEYBEES

Diseases, Parasites, Pests, Predators and their Management

# HONEYBEES

Diseases, Parasites, Pests, Predators
and their Management

**N. NAGARAJA**

Lecturer
UGC Academic Staff College
Central College Campus
Bangalore University
Bangalore

**D. RAJAGOPAL**

Former Director of Instruction
(Post Graduate Studies)
University of Agricultural Sciences
GKVK, Bangalore

**MJP PUBLISHERS**

Cataloguing-in-Publication    Data

Nagaraja, N. (1969 - ).
    Honeybees: Diseases, Parasites, Pests, Predators and
    their Management / by N. Nagaraja and D. Rajagopal. -
Chennai : MJP Publishers, 2009
    xiv, 210 p. ; 23 cm.
    Includes References and Index.
    ISBN 978-81-8094-059-0 (pbk.)
    1. Honeybees, diseases 2. Apiculture, diseases
    3. Diseases, Honeybees-management 4. Management;
    Honeybees-Diseases I. Rajagopal, D II. Title.
    638.154: 658 dc 22 NAG MJP 057

ISBN 978-81-8094-059-0      **MJP PUBLISHERS**
© Publishers, 2009          47, Nallathambi Street
All rights reserved         Triplicane
Printed and bound in India  Chennai 600 005

             Publisher  :  J.C. Pillai
     Managing Editor  :  C. Sajeesh Kumar
       Project Editor  :  P. Parvath Radha
  Acquisitions Editor  :  C. Janarthanan
       Editoral Team  :  B. Ramalakshmi, N. Pushpa Bharathi,
                         L. Mohanapriya, M. Gnanasoundari
                         Lissy John, N. Yamuna Devi,
            CIP Data  :  Prof. K. Hariharan, Librarian
                         RKM Vivekananda College, Chennai.

# FOREWORD

There are many books on honeybee diseases and pests written in different languages. They are mostly concerned with *Apis mellifera* and do not describe or even mention shortly about diseases and pests of other *Apis* species. In Asia there are nine species of *Apis*, which suffer from different diseases and pests.

Several attempts were made to introduce *A. mellifera* into southeast Asia. However, all of them failed due to invasion of the parasitic mites, *Varroa destructor* and *Tropilaelaps clareae*. Only after biology of those mites was studied and control methods were elaborated, successful introduction of *A. mellifera* into southeast Asia was made. After *V. destructor* was introduced to Europe, biology of this mite was intensively investigated and different control methods were elaborated. It was suggested that the biology of *T. clareae* is similar to that of *V. destructor* and similar control methods should be applied. However, after Laigo and Morse (1969) applied twelve weekly fumigations with Folbex (Chlorobenzilate), the population of *V. destructor* was practically eliminated, but *Tropilaelaps* population consistently remained. Similarly, six weekly treatments of Folbex by Atwal and Goyal (1971) were ineffective, although these treatments could control *V. destructor*.

After extensive studies, it was found that the biology of *T. clareae* differs from that of *V. destructor*. A very effective method was elaborated to control *T. clareae* without any medicine. Intensive efforts are being made to breed *A. mellifera* with highly developed hygienic behaviour. Bees, which quickly detect and remove the brood killed by diseases or parasites, are considered as resistant. However, the Asian *Apis dorsata* and *Apis laboriosa* show the opposite mechanism. They do not open and remove dead brood inside sealed comb cells. This way the pathogens and parasites are arrested within capped brood cells and this prevents infection of remaining brood in the colony during the cleaning process. Thus, this method is more effective in preventing the spread of diseases and parasites.

All the phenomena specific to bee diseases and pests in nature, confined to many parts of the world more specifically to southeast Asian regions are described in this book. Dr. N. Nagaraja and Dr. D. Rajagopal have first described briefly the biology of the nine *Apis* species and a species of stingless bee followed by the viral diseases especially the Thai sac brood virus disease, which caused severe damage to *Apis cerana* beekeeping in India. Then they focus on the bacterial diseases like American and European foul brood and also the microsporadian disease, nosemosis. Detailed description on the biology and control methods of the parasitic mites which cause severe threat to *A. mellifera* beekeeping industry are presented. The fluctuation of diseases and pests in specific to southeast Asian conditions, which differ from those in other continents, are being elaborated in detail. Detailed information on natural enemies of honeybees and methods of their management are highlighted by compiling several research findings from different sources. An extensive list of original source of references is included, which are difficult to access not only for practical beekeepers but also for many bee scientists.

Therefore, this book, which has gathered an accessible knowledge, no doubt would serve at first place for beekeepers in southeast Asia, and also for beekeepers and bee scientists all over the world.

**Dr. Jerzy Woyke**
Professor Emeritus, Apiculture Division
Agriculture University, Warsaw, Poland

# PREFACE

Honeybees are vulnerable to a variety of natural enemies, from viral diseases to vertebrate predators, in spite of their highly evolved defensive strategies. Numerous diseases have been reported on both brood as well as on adult bees, but a few would cause catastrophic loss to the beekeeping industry. Equally, parasitic mites, wax moths, ants, wasps and birds cause severe damage to honeybee colonies almost throughout the world.

Extensive information is available on biology, symptoms and management of natural enemies of western honeybee, *Apis mellifera*. Many eastern honeybee species, which were almost free from most of the diseases and pests, suffer from a few viral diseases and pests to a greater extent. The techniques of management of natural enemies of honeybees are spread over in innumerable journals worldwide but only a few books have been published on diseases and pests of *A. mellifera*. An attempt has been made to provide an up-to-date information on natural enemies of all species of honeybees.

This book with a foreword by Prof. Jerzy Woyke, a renowned bee scientist from the Agriculture University, Warsaw, Poland, contains 11 chapters. The first six chapters give a comprehensive information on the biology of honeybees, symptoms, diagnosis and management of diseases caused by viruses, bacteria, fungi and protozoa on both brood as well as on adult bees. Extensive information is being documented on the major bee diseases and their management strategies through manipulative methods, sterilization of combs and equipment, selective breeding and chemotherapy. Chapter seven briefly highlights the different non-infectious disorders, followed by chapter eight on the biology, nature of damage and possible control measures of parasitic mites.

Chapters nine and ten describe the major insect pests and vertebrate predators. Finally, the book concludes with the chapter discussing the future strategies on safeguarding of honeybee colonies for honey production and crop pollination.

We hope this book would be a reference document to students, researchers and faculty members of Zoology, Entomology, Apiculture and Microbiology departments, beekeeping extension personnel and beekeepers across the world in general and southeast Asia in particular. Obviously, it would serve as a unique reference book on enemies of honeybees and their management to the undergraduate and postgraduate students and teachers of Entomology and Apiculture in most of the universities in general and agricultural universities in particular.

We are indebted to Prof. J. Woyke for his technical suggestions on the manuscript and for writing a foreword to this book. The critical comments and suggestions on the manuscript by Dr. Dorothea Brückner, Associate Professor, Department of Biology and Chemistry, University of Bremen, Germany, is gratefully acknowledged. We are thankful to Dr. N. Sreenivasa, Senior Acarologist, Department of Entomology, Dr. Nazeer Ahmed Khan, Professor and Head, Dr. N.G. Ravichandra, Associate Professor, Department of Plant Pathology, Dr. A.K. Chakravarthy, Professor of Entomology, Sri. B.V. Ramakrishna, Coordinator, Production Technology for Gherkins, University of Agricultural Sciences, Bangalore, and Dr. T.N. Raju, BES College of Education, Bangalore, India, for their valuable suggestions in the improvement of the manuscript.

We are also grateful to the Council of Scientific and Industrial Research (CSIR), New Delhi, Departments of Ecology and Environment and Industries and Commerce, Government of Karnataka, Bangalore, India, for providing financial assistance to conduct investigations on management of parasitic mites and Thai sac brood virus disease of honeybees at the University of Agricultural Sciences, Bangalore, which has helped us to incorporate most of the experimental results on natural enemies of honeybees and their management.

The first author expresses his heartfelt gratitude to Prof. C. Chandrasekhara Reddy, Former Chairman, Department of Zoology, Bangalore University, Bangalore, India, for his inspiration and encouragement. He also appreciates his wife Mrs. H.K. Nalina and son Master Akshay for their patience and cooperation in the timely completion of the book.

N. Nagaraja
D. Rajagopal

# CONTENTS

# INTRODUCTION

Man maintained an intimate association with honeybees from time immemorial. Honeybees live in highly organized societies that compete our own in complexity and succeed in dealing with varied challenges posed by the social life, including communication, swarming, nest architecture, etc. They are capable of detecting the toxins as autonomous biosensors in the environment (Bromenshenk *et al.*, 1985).

Honeybees produce honey, the first biological sweet commonly known as *Nature's Golden Wonder*, especially from the nectar of blossoms which they collect, combine with their secretions, transform, and store in combs. Honey is being used through the ages not only for food but also as a medicine in curing various human ailments. It is known to possess special components especially in maintaining vigour, vitality and enhancing longevity of human beings. The beeswax, royal jelly, bee pollen and bee venom are the other major beehive products. Bees are efficient pollinators which play an important role in increasing crop production. Cross pollination by bees not only enhances crop yields, but also benefits even many self-pollinated crops (Rajagopal *et al.*, 1999; Rajagopal and Nagaraja, 2003; Tchuenguem, 2002).

Honeybee colonies in general are vulnerable to various natural enemies for brood, adult bees and hive products. Beekeepers are facing the problem of diseases and pests in honeybee colonies since centuries. During Roman times, though beekeepers could not recognize the causal agents, they were able to identify bee colonies suffering from health hazards. Aristotle (384–322 BC) described several disorders in honeybees and attributed these as intoxication by faulty nectar or pollen. He states in his book *Historia Animalium* that bees suffer from diseases when flowers of the trees are infected with rust in dry seasons. The Bible describes the ways of attack on bee nests by various mammals. In 1586, the German apiculturist, Nickel Jacob noticed certain bee diseases and suggested few methods of their management. Later, Dzierzon (1882) identified foul brood diseases in honeybees. Dadant (1890) thought that dysentery in honeybees was caused by dilution in honey. However beekeepers were following traditional methods to safeguard the bee colonies against natural enemies during the early part of nineteenth century.

The early detection on the occurrence of diseases and attack of pests was rather difficult in the feral colonies of domesticated bees which normally nest in darker cavities with narrow hive entrance and thereby prevent direct observation of brood and adult bees inside the nest. The use of movable frame hives made it possible for the timely inspection of the bee colonies and for moving the combs within and between the bee colonies as and when required. Though such observations are pre-requisite in modern beekeeping practices, it would become a route for faster spread of bee diseases and pests. Similarly, transport of bee

colonies by road, train, ship and air resulted in faster spread of their enemies across the globe.

The level of damage caused by enemies in bee colonies varies with geographical regions. Among bee diseases, only a few viral, bacterial, fungal and protozoan diseases are catastrophic to beekeeping industry throughout the world. The sac brood disease was first recorded at USA (White, 1917) and is widely distributed throughout the world in *Apis mellifera* colonies (Ellis and Munn, 2005). Similarly, Thai sac brood virus that was noticed initially in Thailand during 1976 has caused greater havoc to *A. cerana* beekeeping industry in many Asian countries during 1990s. However, American and European foul brood diseases become active in *A. mellifera* rather than in *A. cerana*. The fungal disease, chalk brood, and the microsporidian disease, nosemosis, are also equally destructive in certain seasons. Many of the protozoan parasites also act as vectors in the transmission of viral and bacterial diseases in bee colonies.

Bees are known to build nests and are the hosts to a variety of parasitic mites (Martin, 1994b). Most of the mites associated with bee nests are saprophagous, but a few species have been evolved to become parasitic. *Acarapis woodi* is the endoparasitic mite, which infests the tracheae of the adult bees. *Varroa jacobsoni* which was recorded in the drone larvae of *A. cerana* for the first time in 1904 (Oudemans, 1904) is widely distributed in most parts of the world with more than one species. *Tropilaelaps clareae* was reported from the Philippines on *Apis dorsata* that has been infesting bee brood with greater parasitism on *A. mellifera*. Similarly, many varieties of new natural enemies are invading the bee colonies from time to time throughout the world. In addition, the *Acarapis woodi* and *Varroa* spp. are known to transmit many viral and bacterial diseases of honeybees.

Wax moths are the major pests of honeybees in tropical and sub-tropical regions. They cause greater damage especially during dearth seasons. The pest which destroys the bee combs in the colonies of *A. cerana*, also initiates absconding in *A. dorsata* colonies. The small hive beetle, *Aethina tumida,* is a destructive pest of *A. mellifera* in tropical and sub-tropical regions of Africa, USA and Australia. Similar to wax moths, it infests weaker bee colonies. Many species of ants and wasps are known to predate on the brood and adult honeybees. Several species of birds and mammals including monkeys play an important role on beekeeping industry in many parts of the world. Occasionally, some of the non-infectious disorders like bee poisoning, chilled brood, colony collapse disorder, etc. may cause partial threat to the bee colonies. Many laborious traditional methods were followed to combat the pests and diseases in honeybee colonies during ancient times.

The beekeepers maintain bee colonies strong and healthy by following suitable beekeeping management practices on the incidence of diseases. Based on the investigations made on the epidemiology of the pathogens and the behaviour of the pests, various effective management strategies have been developed. Large-scale fumigation of the hive parts especially combs have been adopted for inactivating the pathogens. Fumigation using ethylene oxide is effective in eliminating the American foul brood, chalk brood, nosemosis and wax moth damage in bee colonies.

Irrespective of these, successful, least expensive and eco-friendly management of natural enemies is indispensable for overall development of beekeeping, not only for honey and wax production but also for greater crop production through cross-pollination by different species of honeybees.

# Biology of Honeybees

Bees in general belong to the family Apidae under the Order Hymenoptera which comprises of Apini (honeybees), Euglossini (orchid bees), Bombini (bumble bees) and Meliponini (stingless bees). These groups would share many characteristics of which nesting in the cavities seems to be an example of ancestral features, that most Apidae have retained (Koeniger, 1995). The tribe Apini has only one genus, *Apis,* with nine species.

## SPECIES OF HONEYBEES

Among the nine species of the genus *Apis,* red dwarf honeybee, *Apis florea* Fab., common giant honeybee, *Apis dorsata* Fab., eastern honeybee, *Apis cerana* Fab., and western honeybee, *Apis mellifera* Linn. are the major species (Figure 2.1). Other known species are black dwarf honeybee, *Apis andreniformis* Smith, giant mountain honeybee, *Apis laboriosa* Smith, red honeybee, *Apis koschevnikovi* Enderlein, mountain honeybee, *Apis nuluensis* Tingek, Koeniger and Koeniger and Sulawesian honeybee, *Apis nigrocincta* Smith.

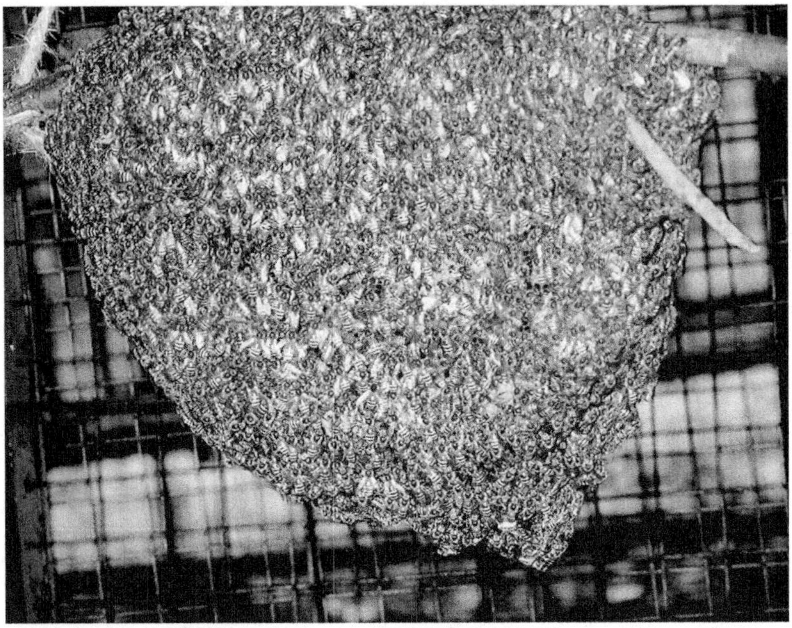

(a)

(b)

(c)

(d)

**Figure 2.1** Major honeybee species (a) *Apis florea* (b) *Apis dorsata* (c) *Apis cerana* (d) *Apis mellifera*

The worldwide distributed *A. mellifera* and Asian distributed *A. cerana, A. koschevnikovi, A. nuluensis* and *A. nigrocincta* are domesticated in man-made beehives. These bees are nesting in cavities where they construct multiple parallel combs, whilst, *A. florea, A. andreniformis, A. dorsata* and *A. laboriosa* have remained as wild honeybees in nature. These are open nesting species, as these invariably build exposed combs.

*Apis florea* found its ecological niche in the stratum of dense bushes and small trees of tropical Asia. These colonies secure shade and live in plains and rarely in places of higher than 1500 m above MSL. They build combs, often suspended from branches of hedges, house chimneys, empty caves and piles of dried sticks. The ability of this species to survive in dry climates is one of the outstanding traits (Ruttner, 1987). This species migrates periodically between plains and adjacent lower hills according to variation in availability of food by deserting old combs and rapidly building new ones (Nagaraja and Rajagopal, 1999). *A. andreniformis* is distributed from Palawan (Philippines) to China and Myanmar. Both *A. florea* and *A. andreniformis* are found to overlap in southeast Asia. These species are more or less similar to each other but they were not generally recognized as separate species. However Wu and Kuang (1987) retained the name as *A. andreniformis* for black dwarf honeybee.

*Apis dorsata* is well distributed in southeast Asian countries both in plains and hilly regions of up to 1600 m. These colonies are highly migratory and normally move to hills during the dry season and return to the plains during the monsoon and post-monsoon seasons, where the flora is quite abundant (Nagaraja and Rajagopal, 2000; Woyke *et al.*, 2005). The colonies are found on the branches of trees, rocks and ceilings of buildings that are normally protected

from direct sun rays and rainfall. The congregation of these colonies is very common and the number ranges from 50 to 200 on a tree or rock where bee forage sources are abundant (Rajagopal and Nagaraja, 1999).

*Apis laboriosa* is found to occur exclusively in mountainous regions of up to 1200 m where night temperatures frequently fall below the freezing point (Roubik *et al.,* 1985). It is well distributed in Uttar Pradesh in India through Nepal, Bhutan, China and northern Laos. It builds nests on cliffs and steep rock faces, frequently in dense aggregations. In spring, the nests are established by migrating swarms at altitudes between 1200 m and 3500 m. In addition, two more subspecies of giant honeybees have been described, i.e., *Apis dorsata breviligula* Maa and *Apis dorsata binghami* Cockerell. The former is found in northwest of the Merrill line in Luzon, the Philippines, whereas the latter is found in the east of the Wallace line in Sulawesi and Butung of Indonesia. *Apis dorsata breviligula* is slightly smaller than *A. d. binghami* but has broader abdomen and shorter mouthparts (Maa, 1953).

Beekeeping with eastern honeybee, *A. cerana*, has been practised in Asia since many centuries. This species is gentle in temperament, easy to handle and industrious in pollen and nectar collection. However, it exhibits frequent swarming and absconding of colonies. It occurs in Asia, east of Iran and south of the great mountain regions and central deserts. Feral bee colonies are generally found in hollows of trees, clefts in rocks, termite mounds, crevices of walls, telephone poles, etc.

The taxonomic studies on cavity-nesting honeybees have confirmed three more distinct species. They are *A. koschevnikovi* (Tingek *et al.,* 1988), *A. nigrocincta* (Hadisoesilo *et al.,* 1996) and *A. nuluensis* (Tingek *et al.,* 1996).

*A. koschevnikovi* is well distributed in the regions of Java, Sumatra, peninsular Malaysia and southern Thailand. It is relatively larger than *A. cerana* and is reddish yellow in colour. These bees differ in a few morphometric characters and also in mitochondrial and nuclear DNA composition from their sister honeybee species. However, their phylogenetic relationships are yet to be established (Arias and Sheppard, 2005). *A. nigrocincta* is well distributed in Sulawesi and Mindanao regions. The drones of these bees perform mating flights about 2 hours later than that of *A. cerana* with a partial overlapping. *A. nuluensis* is confined to mountainous regions at 1500 m and has been known to occur at Sabah in Borneo. The phylogenetical studies based on DNA analysis indicates that *A. nuluensis* and *A. cerana* are distantly related to each other (Arias *et al.*, 1996; Tanaka *et al.*, 2001).

The western honeybee, *A. mellifera,* is the prime domesticated bee, widely distributed all over the world mostly on its introduction from parts of Europe as a good honey producer and efficient pollinator. It has more than 24 subspecies in different geographical regions of Europe and Africa (Ruttner *et al.,* 1989). These bees are gentle and highly industrious with less swarming and absconding tendency, and are well adapted to wider environmental conditions (Nagaraja, 1998a). Among these, one of the subspecies *A. m. ligustica* has been well popularized in many countries across the world. The introduction of *A. mellifera* into tropical Asia was attempted in the late nineteenth century. But, its successful introduction during the 1960s has popularized it as a good honey producer especially in northern and northeastern states of India (Atwal and Goyal, 1973).

In addition to *Apis,* a few species of stingless bees produce honey in their colonies. The stingless bees, which are social, live in different nesting system with permanent colonies

across the tropical world. They do not form clusters like *Apis* spp. The dammer bees, *Melipona (Trigona) iridipennis* Smith are widely distributed in the tropical and sub-tropical regions of India. They build nests in the ground, hollows of trees, bamboo, cracks and crevices in walls of buildings, culverts, bridges, etc. The nests are made up of wax, soil and plant resins. The pollen pots, honey pots and brood cells of the colony are differentiated based on their size (Figure 2.2). The involucrum surrounding the brood cells helps to maintain nest temperature. These bees take care of their brood through mass provision. This species is considered to be the chief pollinator in certain orchard crops rather than as a honey producer as it yields limited quantity of honey. They exhibit defensive mechanism against the enemies by biting through their mandibles.

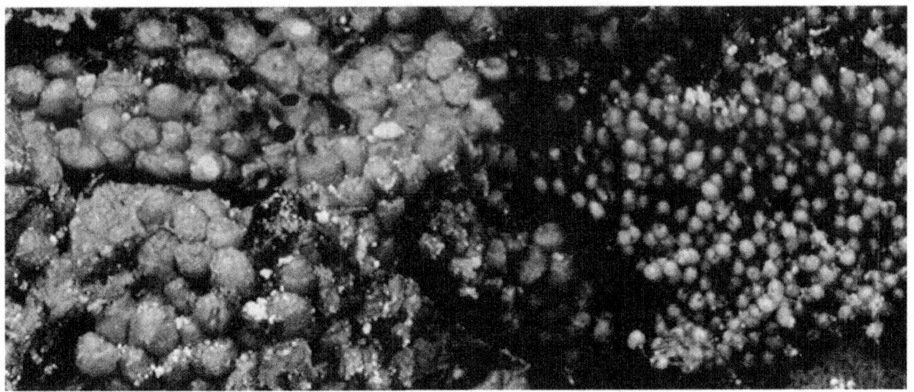

**Figure 2.2** The nest with different cells of *Trigona iridipennis*

## LIFE CYCLE

Honeybee colony normally consists of a fertile queen, hundreds of drones and twenty to fifty thousand worker bees during honey flow season. A honeybee, during its course of development, passes through four metamorphic stages, i.e., egg, larva, pupa and adult. The worker bee develops into an adult from the fertilized egg. The worker brood occupies most of the brood nest in the central part of the comb, which is the most protected and thermally regulated part in the hive. The embryo develops in the egg by absorbing nutrients present in the yolk. On hatching, the worker larvae are fed initially with worker jelly for about two days followed by modified worker jelly added with pollen and honey. Ribbands (1953) opined that the presence of pollen might be accidental in the larval jelly. Larval feeding ceases on attaining the age of about 5 days when the cells are sealed by wax capping. After maturation, the pupa emerges as an adult in about 21 days, though the development duration varies with different species of honeybees (Table 2.1).

Drones develop from unfertilized eggs laid generally by the queen and are fed with drone jelly initially for one to three days followed by modified drone jelly added with honey and pollen. Matsuka *et al.* (1973) found that bees feed modified drone jelly to the male larva for 5 to 6 days. The developmental period of drones from egg to adult is about 24 days which is comparatively longer than that of worker and queen bees. The chief function of the drones is to mate with the virgin queen which on maturation flies out generally in the afternoon to drone congregation areas where mating occurs.

The queen is the mother of the colony, which lays eggs and coordinates all the activities of colony through its glandular secretions, primarily the queen substance. The queen develops from the fertilized egg to the stage of an adult in 16 days. It is fed with royal jelly secreted by the nurse bees. Traces of pollen have been recorded in the food of queen larvae (Ribbands, 1953). It mates with many drones only once in its lifetime. It is known to store nearly five million sperms in the spermatheca and lay millions of eggs during its lifespan of about 3 to 5 years.

**Table 2.1** Duration of brood development (days) in different castes of major honeybee species

| Bee species | Worker | | | | Queen | | | | Drone | | | | Re |
|---|---|---|---|---|---|---|---|---|---|---|---|---|---|
| | Egg | Larva | Pupa | Total | Egg | Larva | Pupa | Total | Egg | Larva | Pupa | Total | |
| *Apis dorsata* | 2.9 | 4.6 | 10.9 | 18.4 | 2.0 | 4.5 | 7.0 | 13.5 | 2.9 | 4.6 | 14.3 | 21.8 | Qa Na |
| *Apis florea* | 3.0 | 6.4 | 11.2 | 20.6 | 3.0 | 6.8 | 7.7 | 17.5 | 3.0 | 6.7 | 12.8 | 22.5 | Sai Sir |
| *Apis cerana* | 3.0 | 4.5 | 11.5 | 19.0 | 3.0 | 5.0 | 7.5 | 15.5 | 3.0 | 7.0 | 13.5 | 23.5 | Su anc (19 |
| *Apis mellifera* | 3.0 | 6.0 | 12.0 | 21.0 | 3.0 | 5.0 | 6.0 | 14.0 | 3.0 | 7.0 | 14.0 | 24.0 | Na Ra |

Worker bees recognize the acute sense of odour in the colony through their antennae. They constantly touch the antennae of their nestmates to solicit food by positioning their mouthparts. The antennation is also helpful in receiving and transferring the chemical information in identifying the bees of their own colony and also their intruders. The chief cleansing structures of worker bees are located on the legs and mouthparts. The forelegs are located on the prothorax close to the head. The bee uses hairs on its basitarsus to clean the dust, pollen and other foreign bodies from the head region. The middle legs have no special tools but hairs covering the inner side of the basitarsus are used for removing pollen from the thorax. The hind legs have highly modified structures known as corbiculae on the tibia and basitarsus which are used to carry pollen load from flowers to the hive. The worker bee uses proboscis in the process of removing pests, parasites and foreign bodies. The sting is a defensive organ in worker bees, and by way of the stinging process that serves to defend the colony against the attack of intruders. However, the sting is longer than worker in queen but vestigial in drones (Crane, 1990).

## ACTIVITIES OF WORKER BEES

Worker bees perform their routine activities within and outside the hive in relation to their age. Most of the duties within the hive are socially organized though they vary according to the secretions of the glandular system. Generally the younger bees tend to stay inside the hive for the first two to three weeks and then involve in outdoor activities for the remaining period of their adult life. The young bees typically clean the cells from which bees have emerged. They begin to feed a mixture of honey and pollen to larvae older than three days. They perform orientation flights also called as "play flights" near the hive

entrance after two weeks and perform duties such as removing debris and dead bees, packing pollen in the cells, comb building, cell capping, nectar ripening and guarding the colony. The house bees on attaining the age of three weeks become field bees and get involved in the collection of nectar, pollen, water and propolis depending on the needs of the colony (Figure 2.3). The field bees collect pollen and nectar

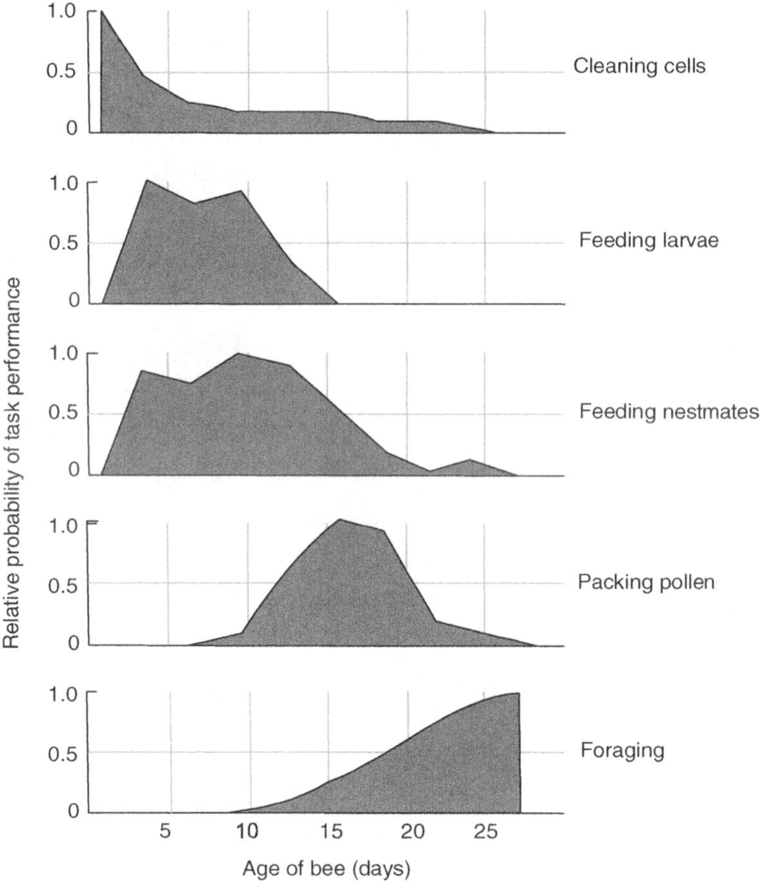


**Figure 2.3**   Age-related activities of worker bees in a honeybee colony (Seeley, 1985)

from the flowers and carry them to the hive in specially modified structures such as the pollen basket and honey sac. The foragers transfer the nectar to house bees on reaching the hive, which later spread the nectar and convert it into honey in the comb cells. However, the pollen is stored in the pollen cells. The nectar and pollen are the common food of the brood and adult bees, and are rich in carbohydrates and proteins, respectively.

## HYGIENIC BEHAVIOUR

Honeybees have long been known to maintain clean and hygienic conditions in their colonies. This social behaviour encompasses uncapping and removal of dead, diseased and infected larvae and pupae from the combs (Woyke *et al.*, 2004). Hygienic behaviour is an ideal trait to determine the disease and pest tolerance in honeybees (Spivak and Gilliam, 1998). Various bioassays have been developed to screen the hygienic colonies. The major assays used are freeze killing brood (Newton *et al.*, 1975; Taber, 1982), pin killing brood (Spivak and Downey, 1998), and freeze killing brood by liquid nitrogen (Reuter and Spivak, 1998). Freeze killed brood in an area of $5 \times 6$ cm comb sections containing about 100 sealed brood cells are introduced into the colonies. These sections are taken out from the combs of the brood frame, frozen at $-20°C$ for 24 hours and replaced in the same frame of the colony. The time required for colonies to detect, uncap and remove the brood is recorded. Colonies that removed the freeze killed brood from the comb section within one or two days are considered as hygienic and colonies that took longer time to remove the brood are considered as non-hygienic (Figure 2.4).

In the pin-killing assay, a fine insect pin is inserted through the sealed brood and the time required for colonies to remove the pierced dead brood is recorded. Since almost a decade, freeze killing by liquid nitrogen is used to freeze a section of comb directly within the frame. A cylinder of thin

(a)

(b)

**Figure 2.4** (a) Hygienic and (b) non-hygienic behaviour in honeybees

metal, 6–8 cm in diameter is formed and twisted into combs containing sealed brood. About 300 ml of liquid nitrogen is poured slowly into the cylinder and the rate of uncapping and removal of dead brood are recorded. Pinkilling technique is less laborious and does not cause much damage to the combs. Although hygienic colonies are selected through different assays to demonstrate their resistance, a highly hygienic colony could be physiologically susceptible and a non-hygienic colony could be physiologically resistant to a disease or pest (Spivak and Gilliam, 1993). Therefore, it is essential to reconfirm the hygienic condition with the pathogens of respective diseases or pests with hygienic breeder stock (Spivak and Gilliam, 1998, Spivak and Reuter, 1998).

## COLONY DEFENCE

Honeybees have developed varied defensive mechanisms against their natural enemies. The wild bees generally construct the nests on the sites free from pests, predators and natural hazards such as rainfall, wind, etc. (Reddy and Reddy, 1993). *A. florea* tends to select dark and shaded sites to protect the colonies against birds and mammals. These bees elicit a hissing behaviour against any disturbance to the colony. The intruder which lands on the comb, is normally attacked by a group of worker bees and sometimes they clump around the intruder (Free and Williams, 1979). However, on continuous disturbance, the colony would rather abscond than defending itself. If an intruder approaches the nest of *A. andreniformis*, a tail of workers hangs beneath the nest by loosening the nest curtain and exhibit shimmering, hissing and roaring behaviour.

*Apis dorsata* builds nests beyond the reach of terrestrial predators. They can be easily alerted to the threat and they

warn the predators by visual and auditory signals. The workers of both *A. dorsata* and *A. laboriosa* twist their bodies in the vertical direction by raising the abdomen and lowering the head whenever such disturbance occurs on their nest (Woyke *et al.*, 2008). *A. dorsata* workers alarmed by disturbed bees, perform quick runs over the nest curtain by forming clusters at the lower edge of the colony. While noting this behaviour, the workers of nearby giant bee colonies also emit defense waves on the surface of the nest (Seeley *et al.*, 1982). This behaviour was also documented in the colonies of *A. laboriosa* (Batra, 1996), *A. andreniformis* (Wongsiri *et al.*, 1997) and *A. florea* (Butler, 1974). Exposure of the sting and opening of the mandibles release the volatile alarm pheromone, which is perceived by nearby colonies and gets them alerted.

The feral colonies of domesticated honeybees have a narrow deep hive entrance and generally maintain hive sanitation by cementing the cracks and crevices with propolis, a resinous substance collected from several species of plants. It provides an important line of defence in these species (Seeley *et al.*, 1982). The defensive mechanism in *A. cerana* consists of shimmering, aggressive and absconding behaviour. Butler (1974) called shimmering as a threatening behaviour in which the bee performs abdominal, side-to-side and left-to-right movements towards the intruder at the hive entrance. *A. cerana* also exhibits defensive response to mechanical stimuli. On a slight knock to the hive, the colony emits a conspicuous hissing sound which is transmitted among the nestmates either by body contact or by a movement produced by the wings.

The tactile contact stimulates shimmering and visual cues in body shaking behaviour in *A. nuluensis* (Koeniger *et al.*, 1996). Domesticated bee species apply blasts of air produced by wing fanning to dislodge insect intruders at the hive entrance. *A. mellifera* collect greater quantity of propolis and pack the cracks and crevices of the hive. Africanized honeybees are generally more aggressive than European honeybees on receiving similar stimulus. These bees identify the avian and mammalian predators by vibratory movements and follow a moving predator for considerable distance when the colony is disturbed. Wager and Breed (2000) found that a moving target without alarm pheromone attracted more number of bees than a stationary source. In addition, juvenile hormone may also stimulate worker bees for colony defence (Pearce *et al.*, 2001). The action of this hormone could be associated with an induction of neural responses by increasing neurotransmitter titers (Schulz *et al.*, 2002).

Colony defence is one of the survival strategies in honeybees to defend against varieties of natural enemies, which are being attracted for a rich source of hive components. The sting is the primary defensive structure and is well developed in worker bees of the age between 12 to 19 days. It bears two barbed lancets and a stylet (Figure 2.5). In a bee that tries to retract the sting after having stung an intruder, the pointed barbs rip the weak membranes attached to sting chamber and the bee dies within a few days. However, a bee that has stung another bee can retract its sting safely. The queen has a larger sting with a few blunter barbs on the lancet and firmly fixed in the sting chamber. She is therefore able to retract her sting after using the same.

The hive entrance is the primary route for the access and entry of intruders into the colony. The basic unit ofcolony defence is an individual worker bee, which guards the colony, identifies and removes non-specific intruders and also recruits nestmates to defend the colony (Breed, 1991).

**Figure 2.5**  Sting and its associated glands in the worker bee of *Apis dorsata*

The colonies with more guard bees are likely to have an effective defensive mechanism where hundreds or even thousands of guard bees involve in pursuing the threat to the nest. The guard bees patrol the colony (Arechavaleta-Velasco *et al.*, 2003) and the special guard bees called as soldiers participate in mass attack against the intruders (Breed *et al.*, 2004).

Guard bees exhibit a sequence of defensive responses against intruders by standing at the hive entrance in a characteristic way by raising their forelegs slightly. The bee's response to the first stimulus (alerting) strengthens its guarding stance by raising its abdomen possibly by protruding sting with antennation. During this process, the guard bee may recruit its nestmates to alert the guard activity by releasing an alarm pheromone. The koschevnikov gland produces most of the alarm pheromonal components, which are volatilized in a rapid way (Cassier *et al.*, 1994). The second stimulus (activating) causes the bees to find out the source of disturbance. On locating the cause, the third stimulus (attracting) initiates the bee to orient itself and move towards the disturbed site. As a result of the fourth stimulus (culminating), it attacks the target intruder by threatening and emitting a high-pitched buzz followed by stinging (Figure 2.6).

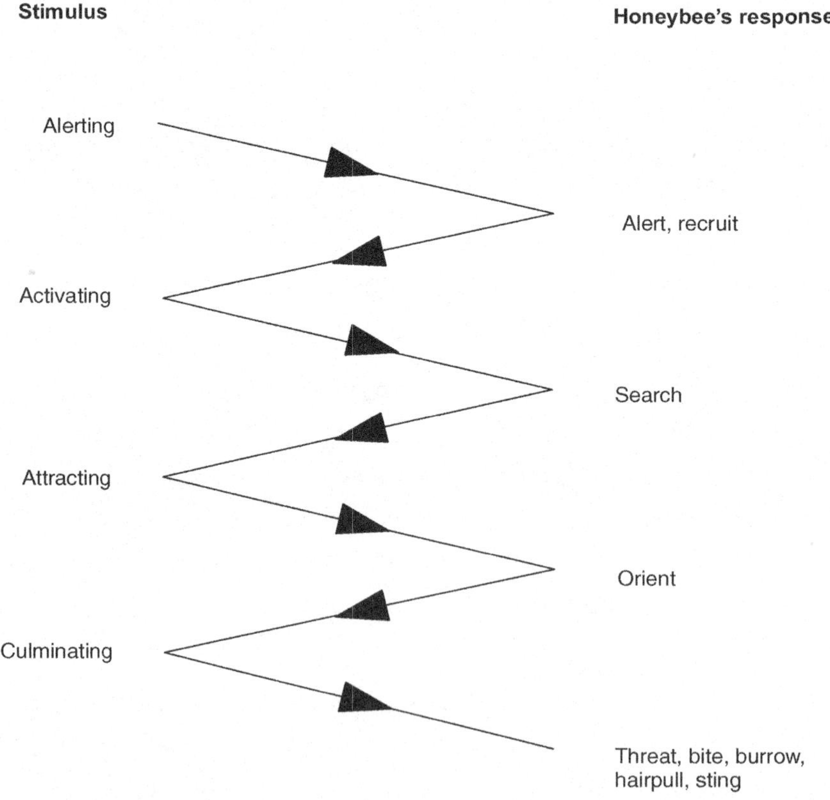


**Figure 2.6**   Sequential defensive behaviour of honeybees (Collins *et al.*, 1980)

**Table 2.2** Major diseases of honeybee species

| Disease | Causal organism | Honeybee species | Reference |
| --- | --- | --- | --- |
| Thai sac brood | *Morator aetatulae* Holmes (Thai strain) | *Apis cerana* | Bailey *et al.* (1982) |
| Sac brood | *Morator aetatulae* Holmes | *Apis mellifera* | Bailey and Ball (1991 |
| Paralysis | Paralysis viruses | *Apis mellifera* | Bailey *et al.* (1983a) |
| Kashmir bee virus | Kashmir bee virus | *Apis cerana* | Bailey and Woods (1! |
|  |  | *Apis mellifera* | Bailey *et al.* (1979) |
| American foul brood | *Paenibacillus larvae larvae* White | *Apis mellifera* | Heyndrickx *et al.* (19 |
|  |  | *Apis cerana* | Singh (1961) |
| European foul brood | *Melissococcus plutonius* Corrig | *Apis mellifera* | Shimanuki (1990) |
|  |  | *Apis cerana* | Kshirsagar and Godb |
|  |  | *Apis laboriosa* | Bailey and Ball (1991 |
| Chalk brood | *Ascosphaera apis* | *Apis mellifera* | Spiltoir and Olive (1! |
|  | (Maassen ex Claussen) Olive and Spiltoir | *Apis cerana* | Gilliam *et al.* (1993) |
| Stone brood | *Aspergillus flavus* Link. | *Apis mellifera* | Alizadeh and Mossad |
| Nosemosis | *Nosema apis* Zander | *Apis mellifera* | Morgenthaler (1963) |
|  |  | *Apis cerana* | Kshirsagar *et al.* (197 |
| Amoeba disease | *Malpighamoeba mellificae* Prell | *Apis mellifera* | Prell (1926) |

Though the guarding behaviour is genetically controlled (Robinson and Page, 1988) strong colonies are likely to express more defensive mechanism than the weaker colonies. The intensity of colony defence can also be attributed to factors such as temperature, humidity, colony size, worker age and availability of food sources in the vicinity (Southwick and Moritz, 1987).

## NATURAL ENEMIES

In spite of several defensive strategies, the honeybee colonies are vulnerable to a variety of natural enemies. These natural enemies are mainly the diseases caused by microbial and non-microbial infections and also pests, predators and parasites. Table 2.2 shows the major diseases, the causal organisms and the species of honeybees affected. Though most of the microbial pathogens kill developing brood, a few cause diseases in adult bees. Brood diseases cause greater damage than adult diseases in honeybee colonies. Honeybee colonies are invaded by a variety of pests for stored hive products, brood and adult bees. Among the pests affecting bee colonies, parasitic mites, greater wax moth, ants and wasps cause severe damage. In open-nesting bee colonies, selection tends to favour the lower virulence of pathogens that do not normally kill the host (Fries and Camazine, 2001).

# 3

# Viral Diseases

Viruses are microscopic entities, whose life processes are apparently dependent upon the metabolism of the host cells. The viruses that cause diseases in insects are similar to most of the other viruses in their basic properties. These are comprised of genomic RNA or DNA bound to a protein coat and are considered to be the smallest and simplest entities capable of multiplication.

The International Committee on Taxonomy of Viruses has placed most of the known viruses into hierarchical levels of family, and in certain cases, sub-family and genus. Viruses that are primarily found on insects are placed under 12 families and an unclassified group (Miller, 1998). Similarly, most of the honeybee viruses are also kept under an unclassified group of RNA viruses.

The studies on insect viruses began with the observations on the diseases of silkworm during the sixteenth century. Interestingly, insects such as *A. mellifera* and *Drosophila melanogaster* Meigen have been subjected to intensive studies.

About 18 viruses have been characterized in honeybees and most of them cause sub-lethal infections (Allen and Ball, 1996). Only a few have been reported in *A. cerana* in contrast to many viruses from *A. mellifera*. However, no virus is reported to cause severe damage to the colonies of *A. dorsata* and *A. florea*.

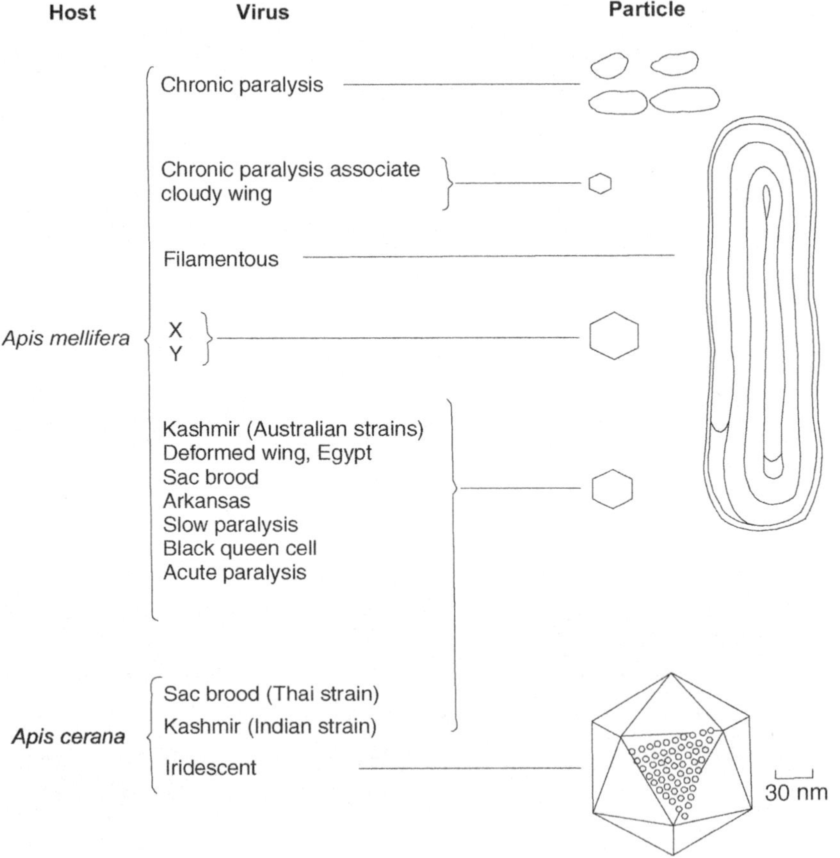


**Figure 3.1**   Diagrammatic outlines of different viruses of honeybees (Bailey and Ball, 1991).

Most of the honeybee viruses are small isohedral particles about 30 nm in diameter. Under the electron microscope, they appear as indefinite spheres and cannot be easily distinguished morphologically. The diagrammatic outlines

of different honeybee viruses are depicted in Figure 3.1. The greater virion sizes are recorded in filamentous and chronic paralysis viruses in contrast to smaller viruses such as chronic paralysis associate and cloudy wing particle viruses. Similarly, viruses X and Y bear proteins with greater molecular weight (Table 3.1). Many viruses have become a threat to honeybee colonies and do not cause significant harm initially but they kill the larvae, pupae and adult bees during the later periods (Anderson, 1995). Honeybee viruses are grouped into viruses infecting bee brood and adult bees but most of the viruses infect both brood as well as adult bees.

## VIRAL DISEASES OF BROOD

## THAI SAC BROOD VIRUS

Thai sac brood virus disease was originated in Thailand during 1976 in the colonies of *A. cerana* and diagnosed initially as sac brood virus of *A. mellifera*. Later, the viral pathogen was isolated and referred to as Thai sac brood virus caused by *Morator aetatulae* Holmes Thai strain (Bailey *et al.*, 1982). Both Thai sac brood and sac brood viruses are distinct and differ in respect of properties such as buoyant density, sedimentation rate and reaction with antisera. Thai sac brood virus disease was spread to India during 1985–86 and caused heavy loss to beekeeping industry by killing over 80 to 90 per cent of *A. cerana* colonies. The expanding phase of beekeeping industry faced a severe setback due to disease and the production of honey and beeswax were drastically decreased. Besides, there was a considerable decline in the production from agricultural, horticultural and plantation crops in the absence of sufficient bee population for cross-pollination. This disease has spread widely on *A. cerana* colonies in many Asian countries.

**Table 3.1**    Characteristics of viruses infecting honeybees (Anderson, 1995, Bailey and Ball, 1991).

| Viruses | Virion size (nm) | $S_{20w}$ (Svedbergs) | Buoyant density in CsCl (g/ml) | Molecular weight of proteins (kD) |
|---|---|---|---|---|
| **RNA viruses** | | | | |
| Thai sac brood virus | 30 | 160 | 1.35 | 30,34,39 |
| Sac brood virus | 30 | 160 | 1.35 | 26,28,31 |
| Black queen cell virus | 30 | 151 | 1.34 | 6,29,32,34 |
| Arkansas virus | 30 | 128 | 1.37 | 41 |
| Egypt virus | 30 | 165 | 1.37 | 25,30,41 |
| Kashmir bee virus (Indian strain) | 30 | 172 | 1.37 | 24,37,41 |
| Kashmir bee virus (Australian strain) | 30 | 172 | 1.37 | 25,33,36 40,44 |
| Chronic paralysis virus | 20×30–60 | 80–130 | 1.33 | 23.5 |
| Chronic paralysis associate virus | 17 | 41 | 1.38 | 15.0 |
| Acute paralysis virus | 30 | 160 | 1.37 | 23,31 |
| Slow paralysis virus | 30 | 176 | 1.37 | 27,29,46 |
| Virus X | 35 | 187 | 1.35 | 52 |
| Virus Y | 35 | 187 | 1.35 | 50 |
| Cloudy wing virus | 17 | 49 | 1.38 | 19.0 |
| Deformed wing virus | 30 | 165 | 1.37 | — |
| **DNA viruses** | | | | |
| Filamentous virus | 150×450 | — | 1.28 | 13.0–17.0 |
| *Apis* iridescent virus | 150 | 2216 | 1.32 | — |

## Symptoms

The combs of healthy colonies typically have a solid and compact brood pattern. Almost every cell from the centre of the comb outward contains egg, larva, or pupal stages. The capping is convex and uniform in colour. However, the Thai sac brood virus-infected sealed brood cells are irregularly scattered on combs with perforations on the capping (Figure 3.2). Thai sac brood virus is confined only to the brood and quite evidently the larvae exhibit disease symptoms. The symptoms are seen in the early larval stage (2–3 days old), and death occurs either in the late larval or in the pre-pupal stage. The dead larvae usually lie at the bottom of the cell with the head typically turned up. The body of the larva changes from creamy white to pale yellow colour, becomes scalelike and attached to one side of the cell at the bottom. Although adult bees of the infected colonies are free from any visible symptoms, they become sluggish and a small proportion of bees goes out for foraging and carries a small quantity of pollen and nectar. During severe infection, the worker bees fail to cope up with the removal of infected brood and colonies will gradually abscond (Rajagopal and Kencharaddi, 2000).

## Diagnosis

Thai sac brood virus-affected larvae turn their heads upward and when lifted with a pointed needle show a saclike appearance (Figure 3.3). Serologically, the virus isolate develops a single, sharp, positive precipitin band against the antiserum of Thai sac brood and sac brood viruses, which showed its closer relation with sac brood virus of *A. mellifera* (Bailey *et al.*, 1982). Since the pathogen is sub-microscopic, the electron microscopic studies of both crude and purified preparations show the presence of isometric particles of 30 nm diameter. Examination

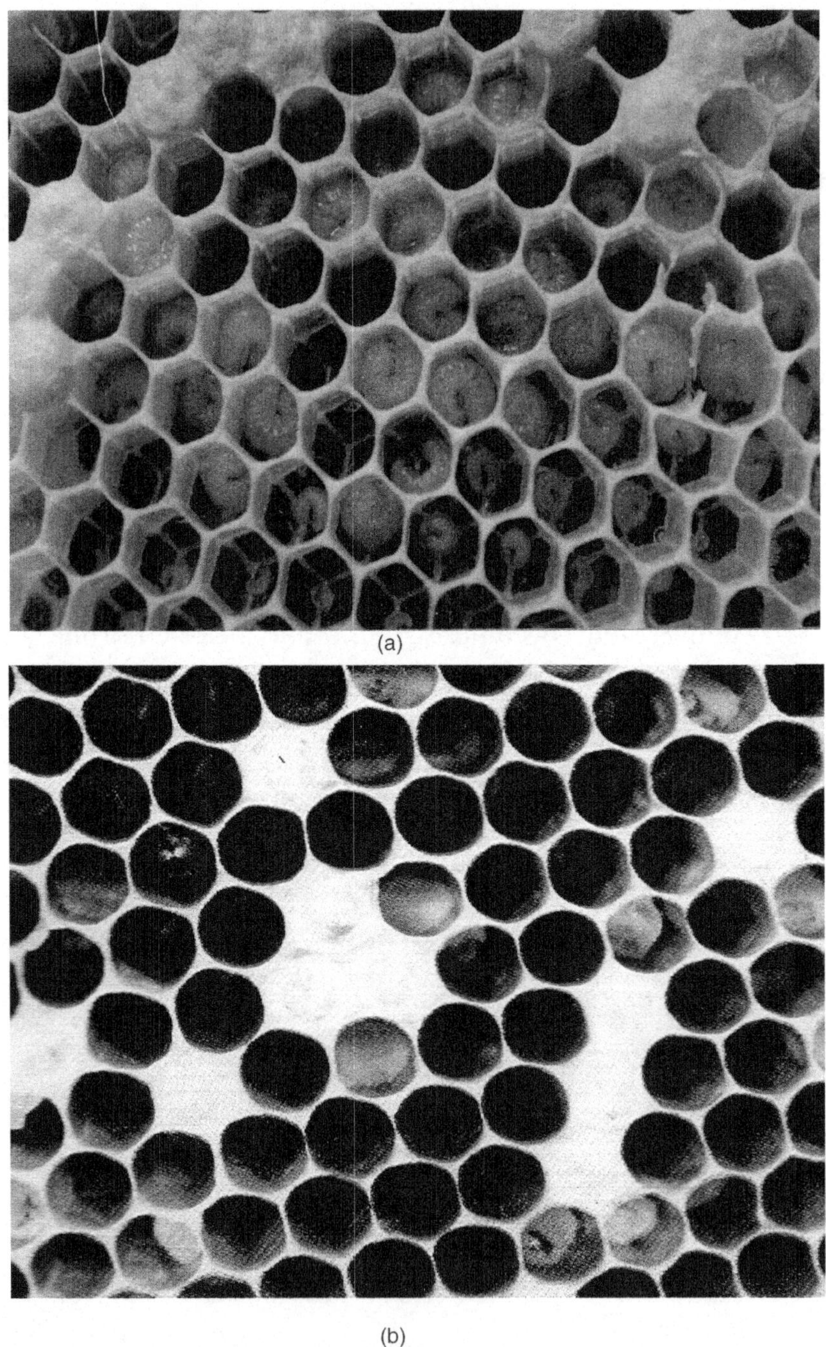

(a)

(b)

**Figure 3.2**   (a) Healthy honeybee larvae   (b) Scalelike dead larvae in TSBV-infected comb

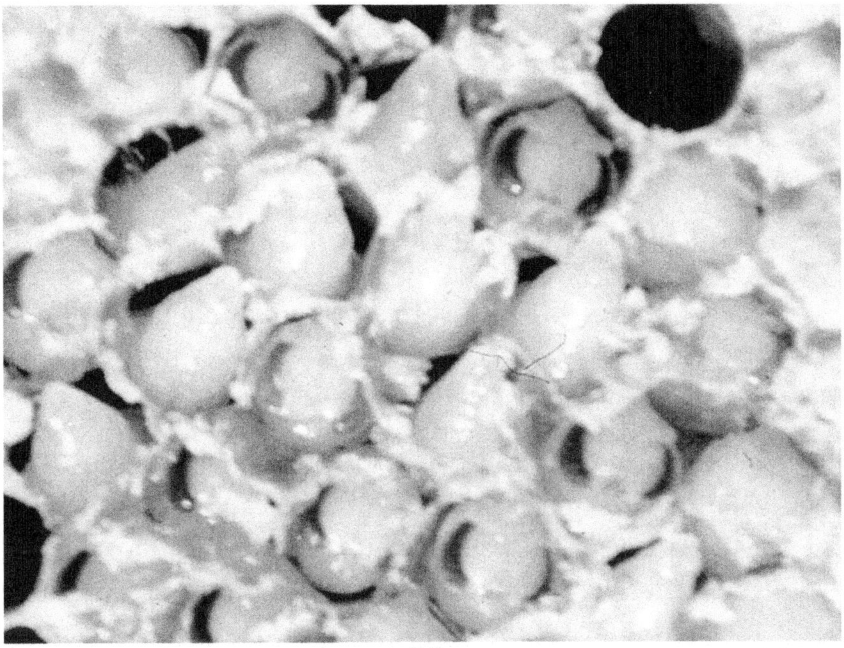

(a)

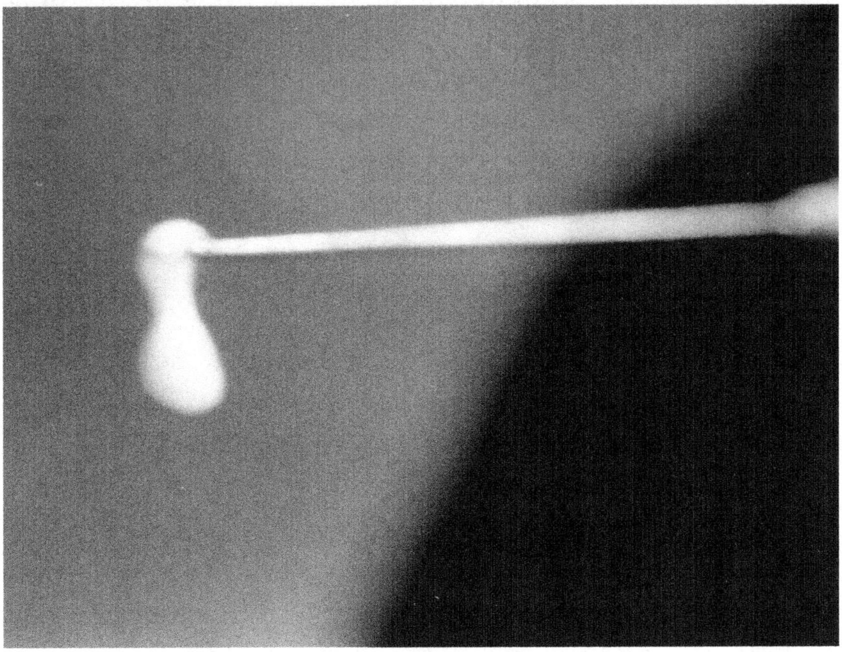

(b)

**Figure 3.3**    (a) Larvae turned their heads upward    (b) Saclike appearance of TSBV-infected dead larva

of ultrathin sections of the midgut of infected adult bees reveals bundles of virions accumulated next to the peritrophic membranes in the gut lumen. The virions are confined to limited areas of cells with disruption of the cytoplasm. The pathogenesis of the virus could be confirmed by feeding the purified virus to the colonies of *A. cerana* which generally express the disease symptoms in a week. However, such symptoms are not noticed in the colonies of *A. mellifera* that are fed with the same virus for several months.

The virus remains accumulated in the hypopharyngeal glands of young adult bees without affecting their health (Bailey, 1969). The infected nurse bees transmit the disease while feeding the larvae with their glandular secretions. It is generally difficult to prevent the migration of the feral infected colonies, which could remain as the primary source of infection. Although deserted colonies leave the infected brood and combs behind, the adults would spread the inoculum through exchange of food and also in the process of cleaning the infected brood cells. Symptoms of Thai sac brood virus disease have also been recorded in the brood of *A. dorsata* and *A. florea*, but the presence of disease has not been confirmed serologically and it is not clear whether bees of these colonies could serve as reservoirs of infection.

Thai sac brood virus disease was prevalent in *A. cerana* colonies during 1992–2002 in Karnataka. The intensity of the disease was more prevalent under stress conditions. The frequent migration of colonies increase the intensity of the disease by spreading the virus inoculum. The disease spreads through contaminated food, overcrowding of colonies, swarming and robbing. It is more common in brood-rearing seasons as the pathogen multiplies in the early brood stage. The intensity of Thai sac brood virus disease is negatively correlated with mean temperature and bright

sunshine hours, whereas the relative humidity is positively correlated (Rajagopal and Kencharaddi, 2000). However, Kannan (1996) found no correlation between climate and susceptibility of brood to the virus. The foraging activity with respect to number of pollen and nectar foragers was comparatively high in healthy colonies than diseased colonies.

*Varroa jacobsoni* has been observed to be associated with brood of honeybees infected with Thai sac brood virus. A survey from five selected districts of Karnataka showed that about 33 to 86 per cent of TSBV-infected *A. cerana* colonies had *Varroa* mites (Table 3.2). These findings confirm the possibility of *Varroa* mites as vectors in spreading the Thai sac brood virus disease in *A. cerana* colonies.

**Table 3.2** Infestation of *Varroa jacobsoni* in Thai sac brood virus-infected and disease-free colonies of *Apis cerana* in Karnataka, India, during 1997 (Hanumanthaswamy, 2000)

| Surveyed locations | No. of colonies observed | Per cent TSBV-infected colonies | | Per cent disease-free colonies | |
|---|---|---|---|---|---|
| | | With mites | Without mites | With mites | Without mites |
| Belgaum | 34 | — | 100.00 | 37.50 | 62.50 |
| Kodagu | 72 | 85.71 | 28.57 | 17.75 | 82.14 |
| Dakshina Kannada | 83 | 66.66 | 33.33 | 31.08 | 68.91 |
| Chikmagalur | 62 | 38.46 | 61.53 | 57.14 | 42.85 |
| Hassan | 44 | 33.33 | 66.66 | 54.28 | 45.71 |

## Management

***Manipulative methods*** Bee colonies are always kept hygienic through periodic cleaning of the hives. On incidence of the viral disease, the entire adult bee population may be transferred to a new or disinfected box supplied with comb foundation sheets. Such colonies are fed with sugar syrup until they construct new combs and store sufficient food. The honey from diseased colonies may not be fed to the healthy colonies as the viral particles may remain in honey for longer periods. During severe infection, the combs containing diseased larvae are burnt to prevent further contamination. Shah and Shah (1988) found that a break in brood rearing either by de-queening or by caging the queen encourage bees to remove infected dead brood efficiently, thereby keeping the infection under check and averting collapse of the colony.

***Breeding for disease tolerance*** *Apis cerana* colonies are known to develop a certain degree of tolerance against Thai sac brood virus disease (Reddy *et al.*, 2004). Such colonies monitor the level of disease through rapid detection and removal of infected dead larvae. The daughter colonies bred from tolerant stock show gradual decline in disease from 50 per cent in parent colony to only 4 per cent in daughter colonies (Verma *et al.*, 1986). Similarly, the yellow strain of *A. cerana* is known to be highly tolerant to disease than the black strain. Systematic breeding of such tolerant colonies is more effective in managing the Thai sac brood virus disease in *A. cerana* colonies. Chinh (2000) obtained tolerant colonies for Thai sac brood virus disease through selective breeding programmes in Vietnam. The feral colonies of *A. cerana* have higher level of survival against the disease than the domesticated colonies perhaps possibly due to the inbreeding in the latter (Varma and Joshi, 1988).

***Chemotherapy*** Antiviral drugs are quite effective against Thai sac brood virus disease. These drugs are either fed to the adult bees with sugar syrup or sprayed on to the infected brood combs at an early stage of infection. Acyclovir

(zovirax), an antiviral compound available in the form of tablet (200 mg), can be dissolved in 100 ml of sugar syrup and fed to the bees at an interval of one week. Similarly, one ml of ribavirin (virazide) is mixed with 100 ml of sugar syrup and fed to the infected bee colonies at an interval of one week. The treatment with acyclovir, ribavirin and homeopathic medicine (antiviral drugs used against viral infections in human beings), gradually reduce the disease infection (Table 3.3). However, under severe infections, ribavirin may not be very effective. Similarly, antiviral drugs like omantadine (100 mg) declined the intensity of the disease after a month of the treatment but, rifamycin (75–150 mg) and cytarbine (0.1–0.2 ml) were not found to be effective as per the results obtained from the series of experiments conducted at the University of Agricultural Sciences, Bangalore, India (Rajagopal and Kencharaddi, 2000).

**Table 3.3**     Efficacy of antiviral drugs against Thai sac brood virus disease at different days after treatment (Rajagopal and Kencharaddi, 2000)

| Treatment | Per cent larval infection | | | | |
|---|---|---|---|---|---|
| | Before treatment | After treatment | | | |
| | | 7 days | 14 days | 21 days | 28 days |
| Acyclovir | 36.42 | 34.20 | 30.20 | 25.66 | 21.28 |
| Ribavirin | 30.68 | 26.21 | 22.28 | 16.82 | 12.50 |
| Homeopathic medicine | 34.56 | 30.26 | 22.82 | 15.66 | 10.22 |
| Untreated check | 32.48 | 36.48 | 42.44 | 48.66 | Deserted |

Leaf extracts of basil, *Ocimum sanctum* Linn., adathoda, *Adathoda vasica* Linn., gripe weed, *Phyllanthus niruri* Linn., clerodendron, *Clerodendron serratum* Linn., wild brinjal, *Solanum indicum* Linn., heart leaf, *Tinospora cardifolia* (Willd.)

Hook. and Thomson, periwinkle, *Catharanthus roseus* (Linn.) G. Don and rhizome extracts of sweet flag, *Acorus calamus* Linn., mango ginger, *Curcuma amada* Roxb. and turmeric, *Curcuma longa* Linn. were attempted in the control of the disease. However, none of the extracts were found to be effective in reducing the intensity of Thai sac brood virus disease.

## SAC BROOD VIRUS

Sac brood virus disease is one of the foremost important viral diseases recorded from *A. mellifera* and is the most widely distributed of all bee viruses infecting brood as well as adult bees (Ellis and Munn, 2005). Similar to Thai sac brood virus, sac brood virus multiplies in tissues of young larvae. Two- to three-day-old larvae are highly susceptible to the disease and the infected larvae fail to pupate but remain stretched on their back by extruding their head towards cell capping (Figure 3.4). The larval cuticle becomes a transparent sac accumulated with a fluid between the epidermal layers. The colour of the infected larvae changes from pearly white to pale yellow and then to dark brown. Within a day after ingestion, the virus begins to multiply in the hypopharyngeal glands of the host (Bailey, 1969). The virus also multiplies in young adult bees without expressing the visible symptoms. But they cease to feed on the pollen as their hypopharyngeal glands undergo degeneration (Du and Zhang, 1985). Adult bees would serve as a source of infection where nurse bees, robbing bees and drones of the diseased colonies may probably transmit the virus. Sac brood virus is reported to be a common virus in Australia but occasionally causes serious loss of brood (Dall, 1985). The ability of the *Varroa* mite in transmitting the virus from infected to healthy colonies has been demonstrated (Ball, 1989).

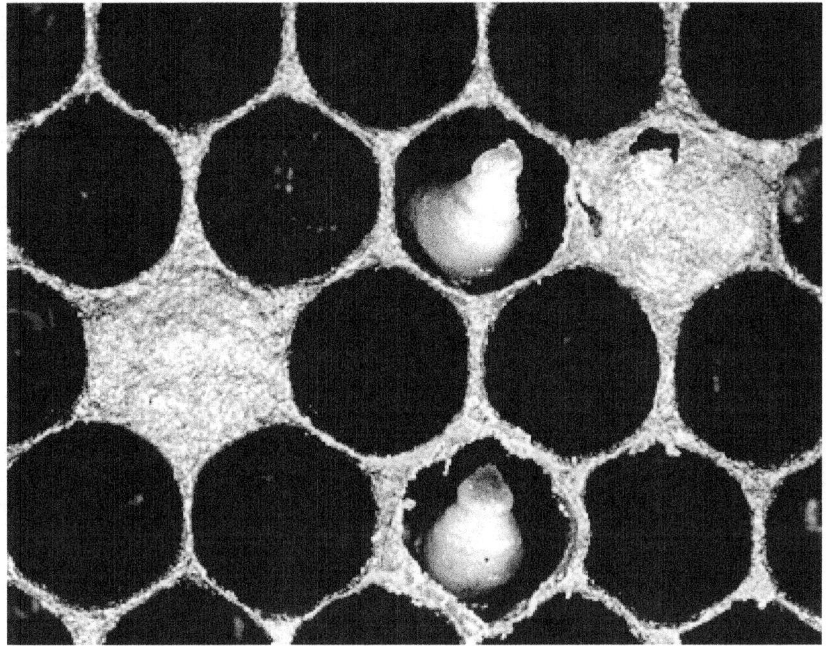

**Figure 3.4**    Sac brood virus-infected larvae with head region upwards in the cell

## BLACK QUEEN CELL VIRUS

Black queen cell virus occurs as virulent in pre-pupal and pupal stages of queen bees of *A. mellifera* (Bailey and Woods, 1977). The infected pre-pupae turn pale yellow in colour and form a tough sac with darker cell walls. Laboratory studies revealed that the virus is dependent on the microsporidian parasite, *Nosema apis,* for the infection of adult queen bee (Bailey *et al.,* 1983b). However, it does not readily multiply on adult worker and drone bees on ingestion.

## VIRAL DISEASES OF ADULT BEES

## ARKANSAS AND EGYPT BEE VIRUSES

Arkansas bee virus is found to infect the bees of *A. mellifera* in Arkansas on injecting into healthy foragers returning to their colonies (Bailey and Woods, 1974). It was also noticed on dead bees infected with chronic paralysis virus. However, Lommell *et al.* (1985) reported that the Californian bees and the original isolates of Arkansas bee virus contained another particle unrelated to any known bee virus. Similarly the Egypt bee virus was isolated from adult bees of *A. mellifera* in Egypt (Bailey *et al.*, 1979).

## KASHMIR BEE VIRUS

Kashmir bee virus is a pathogen of *A. cerana* that killed thousands of colonies in Kashmir (Bailey and Woods, 1977). After introduction of *A. mellifera* into southern Asia, the virus also made a species jump from *A.cerana* and parasitizes the new host, *A. mellifera*. The major symptoms of the disease are gradual weakening of bee colonies with increased number of dead and dying bees near the hive. The infected adult bees are partly or completely hairless, have a dark upper thoracic surface and show trembling uncoordinated movements. Kashmir bee virus multiplies in epithelial tissues of the gut and epidermis. Infected cells show marked pathological changes such as condensation of chromatin, disruption of cytoplasmic organelles and cytoplasmic lobules. It also multiplies in the pupae of *A. mellifera* (Dall, 1987). The high virulence of the pathogen could be possible from the association of *Varroa* mite that transmits the virus into the host's tissues.

Serologically related strains of Kashmir bee virus have been isolated from the adult bees of *A. mellifera* in Australia (Bailey *et al.*, 1979) and New Zealand (Anderson, 1985). It has been noted that the virus enters from stingless bees of the genus *Trigona* that are native to both the countries and southeast Asia. A few strains of Kashmir bee virus have been isolated from dead bees of *A. mellifera* from Canada, USA

and Spain (Bruce *et al.*, 1995). It is proved to be the most virulent of all known honeybee viruses and the injection of only a few viral particles into the haemolymph of adult bees or brood can cause death within three days (Bailey *et al.*, 1979). The virus is possibly transmitted through salivary gland secretions of adult bees in inapparent levels. Characterization of several strains of Kashmir bee virus demonstrates that it is serologically related to acute paralysis virus (Allen and Ball, 1995).

## PARALYSIS VIRUSES

Four types of viruses viz.; chronic paralysis virus, chronic paralysis associate virus, acute paralysis virus and slow paralysis virus are known to cause paralysis in *A. mellifera.*

Chronic paralysis virus is one of the first viruses to be isolated from honeybees (Bailey *et al.*, 1963). The virus multiplies mostly in the head of the adult bees and infects different tissues including the brain and nerve ganglia (Ball *et al.*, 1985). The infected bees become hairless and shiny and have bloated abdomen with partially spread dislocated wings. They exhibit an abnormal trembling motion of the wings and body, fail to fly out and are often seen crawling on the ground and cluster on the top of the hive. It has been detected to be transmitted through bee faeces (Ribiere *et al.*, 2007). Occassionally it kills larvae and pupae in association with the brood mite, *Varroa* (Ball and Allen, 1988).

Chronic paralysis associate virus is one of the smallest viruses known to be associated with the multiplication of chronic paralysis virus. It is a satellite virus, which depends upon cloudy wing virus for replication, although these two viral particles are serologically unrelated (Bailey *et al.*, 1980a). Its prevalence is more evident in queen and worker bees.

Acute paralysis virus has been detected in adult bees, dead queen larvae and *Bombus* spp. in Britain (Bailey and Gibbs, 1964). It is reported in adult bees of *A. mellifera* in France, Italy, Canada, New Zealand and probably from Australia. It is serologically related to Kashmir bee virus and both of these viruses may be a strain of any one of the viruses (Allen and Ball, 1995). The virus was also found in the salivary glands and pollen loads carried out by the bees. Large amount of viruses have been detected in dead bees and brood infested with *Varroa*. *Varroa* is thought to transmit the virus directly from adult bees to pupae and indirectly to larvae by contaminated larval food. This disease is a major cause of mortality in mite-infested bee colonies and it was only honeybee virus that is known to have an alternative host in nature (Ball and Allen, 1988).

Slow paralysis virus was found occasionally in the extracts of adult bees and causes death on injection of virus into the bees. It was originally detected as an inapparent infection in adult bees in association with Bee virus X. It causes mortality of the brood and adult bees in the colonies infested with *V. destructor*.

## FILAMENTOUS VIRUS AND BEE VIRUSES X AND Y

Filamentous virus was first recorded in the haemolymph of adult bees (Clark, 1978). It multiplies in the fat body, ovarian tissues and haemolymph of adult bees. The haemolymph of infected bees is milky white in colour. The virus has a common association with *Nosema apis*. It is the most common and probably the least harmful of all the honeybee viruses (Bailey *et al.*, 1983b).

Bee virus X multiplies in the gut of adult bees. It accelerates the death of bees infected with the protozoan, *Malpighamoeba mellificae*. Bee virus X has been detected in the colonies of *A. mellifera* in all continents except Asia. Since it multiplies slowly, it requires long-lived overwintering bees to establish infection. Bee virus X is distantly related serologically to the bee virus Y. The virus was originally isolated from *A. mellifera* in the United Kingdom. Bee virus Y infects and multiplies in the gut of adult bees in association with *N. apis* without any visible symptoms (Bailey *et al.*, 1980b).

## CLOUDY WING VIRUS

Cloudy wing virus primarily causes infection in adult bees but occasionally in the brood of *A. mellifera* (Anderson, 1993). Infected bees show a marked loss of transparency of wings with a crystalline array of virus particles in muscle fibrils. The virus is airborne and transmitted during congregated colony conditions. It has been recorded in bee colonies infested with *Varroa* mite in Greece and Britain (Allen and Ball, 1996).

## DEFORMED WING VIRUS

Deformed wing virus (DWV) has been classified under RNA virus and is found to be pathogenic to honeybees. The infected bees are characterized by morphological abnormalities including malformed appendages, reduced abdomen and change in colour. Deformed wing virus causes mortality of brood and adult bees of *A. mellifera*. It is serologically related to Egypt bee virus and the infected bees show partially or completely deformed wings. The virus has been detected in the colonies of *A. mellifera* from Europe, Africa, the Middle East and Asia and also on *A. cerana* in China. It was found in association with *Varroa* and the ability of the mite to transmit the virus has been demonstrated (Ball, 1989). It was noticed in the absence of *Varroa* in dead queen and adult bees from Britain and South Africa. It was also recorded in dead bees and brood of *A. mellifera* infested with the mite, *T. clareae* in Nepal (Allen, 1995).

## *APIS* IRIDESCENT VIRUS

Iridovirus was first isolated in *A. cerana* from Kashmir (Bailey and Ball, 1978). It multiplies in the tissues of fat body, alimentary canal, hypopharyngeal glands and ovaries. The tissues of the infected bees become blue-violet to green on illumination with bright white light. This disease is known to reduce the egg-laying capacity of the queen, and the infected worker bees become sluggish and form clusters at the hive entrance and a few may crawl on the ground. This virus was also recorded in the colonies of *A. mellifera* from the northern states of India (Mishra *et al.*,1980).

## OTHER BEE VIRUSES

Berkeley bee virus was isolated from *A. mellifera* in California. In addition, a few more viruses have been reported on honeybee colonies from China. An isometric virus was isolated from the brood of *A. cerana* suffering from the large larval disease. Similarly, the dead melanized pupae of *A. mellifera* contained $42 \times 32$ nm virus particles which have been confirmed as pathogenic from inoculation experiments (Allen and Ball, 1996).

## DETECTION OF BEE VIRUSES

Viruses cannot be cultured on artificial media as they require host-specific living cells for their multiplication. The experimental infection of the virus could provide information on its incubation period, mode of infection and threshold of inoculum. Suitable life stages and the methods of inoculation for cultivation of honeybee viruses have been

determined (Bailey, 1981). Obviously, most of the bee viruses can be multiplied on the bee pupae or adult bees. Majority of the bee viruses are very small and are difficult to distinguish morphologically but the large viruses such as filamentous and *Apis* iridescent viruses can be detected under electron microscope. Physical properties are also useful in differentiating chronic paralysis and cloudy wing viruses.

The outer protective capsid of the virus, which encloses the genetic material is made up of proteins. The techniques of isolation of viruses have been well-documented. Accordingly, the detailed methods of extraction and purification of viruses are discussed. The extraction procedure involves grinding bees in a suitable buffer and solvents and filtering through muslin. Such preparations are confirmed by low and high speed centrifugation. Further purification is usually required to obtain virus preparations that are suitable for antiserum production.

Antibodies against purified preparations of virus particles are readily produced in rabbits by immunizing each with purified virus particles over an eight-week period. Blood samples may be removed from the ear veins of each rabbit about 1 to 2 weeks after the final injection and the serum is separated from clotted blood. This serum, which contains virus-specific antibodies, may be used later in serological tests to detect and identify viruses.

Several serological techniques have been demonstrated for detecting bee viruses (Anderson, 1984). The simplest and most inexpensive test is the immunodiffusion technique. Similarly the enzyme-linked immunosorbent assay and reverse transcriptase (RT) polymerase chain reaction (PCR) (Benjeddou *et al.*, 2001) are also used in detection of viruses (Allen *et al.*, 1986). RT-PCR is effective in detecting

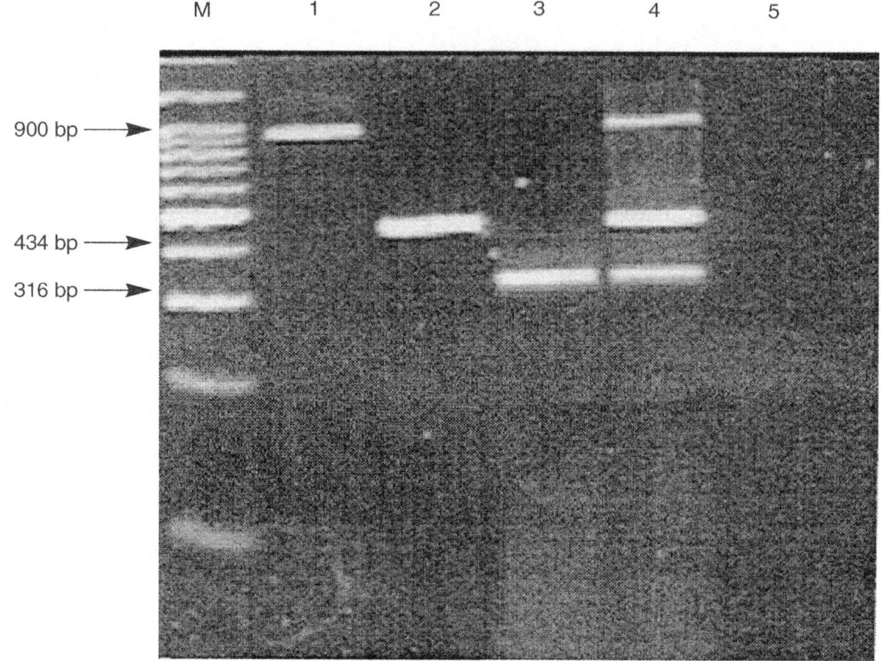

**Figure 3.5**    Amplification of honeybee viruses on the pupae of honeybees individually and by M-RT-PCR, Lane M—100 bp marker; lane 1—ABVP; lane 2—SBV; lane 3— BQCV; lane 4—multiplex of ABPV; SBV and BQCV; lane 5—water (negative control) (Topley *et al.*, 2005).

Kashmir bee virus, sac brood virus, black queen cell virus and acute bee paralysis viruses (Evans and Hung, 2000). Similarly, a single multiplex-RT-PCR assay was developed for the simultaneous detection of three honeybee viruses, viz., acute bee paralysis virus (ABPV), sac brood virus (SBV) and black queen cell virus (BQCV) (Topley *et al.*, 2005). Unique PCR primers were designed from the complete genome sequence to amplify fragments of 900 bp from ABPV, 434 bp from SBV and 316 bp from BQCV (Figure 3.5). Enhanced chemiluminescent western blotting technique uses the antibodies to probe the virus proteins after they are denatured and separated by polyacrylamide gel electrophoresis (PAGE) (Allen *et al.*, 1986).

## MANAGEMENT

A number of manipulative measures would prevent viral diseases in honeybee colonies. Strong and vigorous bee colonies kept in the areas rich in floral sources have generally showed low level of infection. Viruses are probably transmitted through exchange of infected combs and equipment between healthy and infected colonies during routine apiary practices. Avoiding exchange of such infected combs and use of sterilized beekeeping equipment are effective in preventing viral diseases. Similarly, some of the colony manipulative methods used against Thai sac brood virus disease could also be effective in managing sac brood, black queen cell virus and other brood-infecting viral diseases.

The most crucial stage in the dynamics of virus infection is the mode of virus transmission. Transmission of viruses can occur through two pathways, viz., horizontal and vertical. In horizontal transmission, viruses are transmitted among individuals of the same generation, while in vertical transmission, the transmission occurs from adults to their offspring. The horizontal transmission of few viruses in honeybee colonies is primarily associated with different species of vectors. The brood mite, *Varroa destructor* is known to transmit cloudy wing virus, acute paralysis virus, deformed wing virus, sac brood virus, slow paralysis virus, black queen cell virus, Kashmir bee virus, chronic paralysis virus (Chen and Siede, 2007) and Thai sac brood virus. Similarly the parasite, *N. apis,* is associated with the black queen cell virus, bee virus Y, filamentous virus, Kashmir bee virus and chronic paralysis virus and also *M. mellificae* with bee virus X. The brood mite *T. mercedesae* is a biological vector of deformed wing virus, which would open a novel route of virus spread in *A. mellifera* (Dainat *et al.,* 2009).

The small hive beetle, *A. tumida* also acts as a potential biological vector of honeybee viruses (Eyer *et al.*, 2009). Chen *et al.* (2006) reported the vertical transmission of black queen cell virus and deformed wing virus from the queens to offspring by reverse transcription PCR methods. The control measures followed to prevent the transmission of these viruses through the aforesaid vectors would reduce the problem of viral diseases. Some of the viral diseases are more prevalent in the honeybee colonies congregated in large numbers. Therefore, avoiding overcrowding of bee colonies in apiaries reduces the incidence of viral diseases. Mishra *et al.* (1980) fed vitamin B-complex, antibiotics and yeast as a protein source along with sugar syrup to prevent *Apis* iridescent virus disease with only limited success.

# 4

# Bacterial Diseases

Bacteria are ubiquitous, microscopic, prokaryotic organisms infecting a wide variety of fauna and flora in nature. Bacteria infecting insects can be divided into two broad categories as non-spore-forming bacteria and spore-forming bacteria.

American foul brood and European foul brood are highly destructive and widely distributed bacterial diseases of honeybees. Schirach (1771) was the first to use the name foul brood. He apparently used the term to refer to more than one disorder of bees. It was probably Dzierzon (1882) who recognized two kinds of foul brood diseases. A similar view was expressed by Cheshire (1884), but later he opined that there was only one kind of foul brood disease. Phillips (1906) used the terms European and American foul brood to distinguish these two bacterial diseases in honeybees.

## AMERICAN FOUL BROOD

American foul brood is one of the most destructive infectious brood diseases killing millions of *A. mellifera* colonies throughout the world. It is highly contagious and occurs in all seasons on bee brood. In India its occurrence was reported on *A. cerana* colonies from Uttar Pradesh (Singh, 1961).

*Paenibacillus* (formerly *Bacillus*) *larvae* subsp. *larvae* White (Bacillales: Paenibacillaceae) hereafter referred to as *P. larvae*, causes the American foul brood (Heyndrickx *et al.*, 1996). It is a spore-forming, gram-positive and aerobic bacterium. The bacterium is slender, rod-shaped with slightly rounded ends and has a tendency to grow in chains. It varies from 2.5 to 5.0 $\mu$m in length and 0.5 to 0.8 $\mu$m in width. The spores are oval and approximately $0.6 \times 1.3\ \mu$m in size. They can survive on larval food, larval scales and in soil for many years and are able to germinate even after 35 years (Haseman, 1961).

The bacterium occurs in vegetative as well as in the spore form, but the vegetative forms are unable to multiply in the larval intestine (Davidson, 1973). The vegetative cells migrate to the peritropic membrane of the larvae by the process of phagocytosis. Later, they enter the haemocoel and multiply to cause the infection. Old larvae and adult bees may be exposed to the bacterial spores, but infection can be seen only on ingestion of the bacterium into young larvae, 2 to 3 days old. The spores germinate in the lumen of the larval midgut. The strains of *P. larvae* show greater variation in infectivity and some strains require many spores to infect a larva (Genersch *et al.*, 2005).

## Symptoms

Bee colonies infected with American foul brood show non-emergence of few isolated brood cells on the comb. In advanced conditions, the diseased brood is irregularly intermingled with healthy brood with uncapped, punctured or sunken capping in the form of pepper box (Figure 4.1).

The diseased brood is initially dull white in colour and gradually changes to light brown or dark brown. Death of an infected larva usually takes place after the cell is sealed and the cocoon has been spun (White, 1920). Generally, segmentation of the larva is well-marked and gives off fish-gluelike foul odour. When a fine needle is pushed into the remains of the larva and withdrawn, it drops as a semi-fluid in the form of a string. Sometimes the remains dry up and form blackish scales at the bottom of the cell.

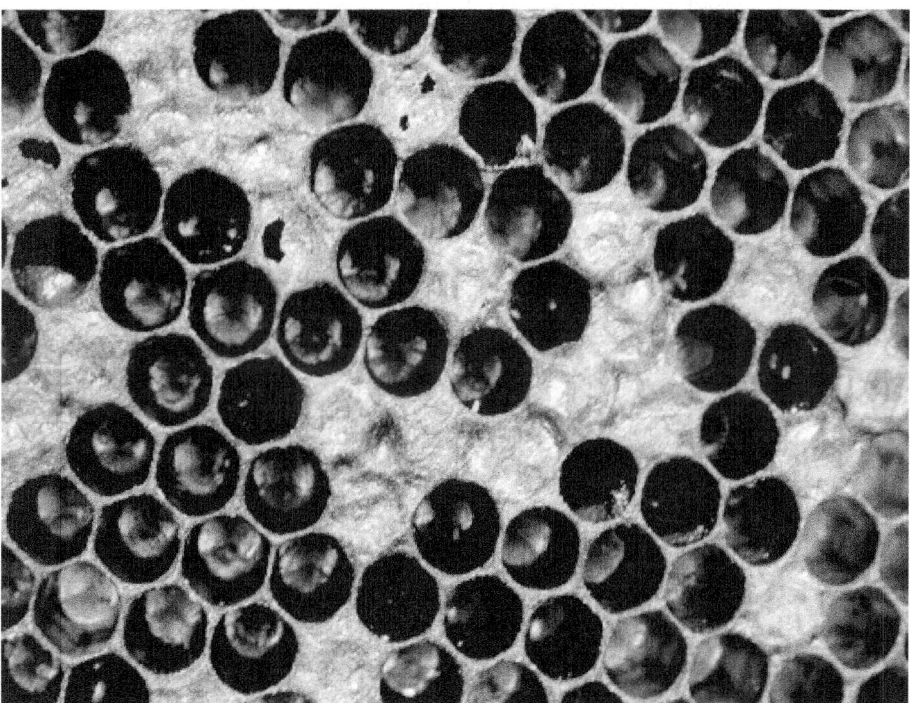

**Figure 4.1**   Honeybee larvae infected with American foul brood

## Diagnosis

The most common method used for detection of the American foul brood is inspection of the brood combs for clinical symptoms. The modified hanging drop technique is effective in differentiating American foul brood from other brood diseases. *P. larvae* spores exhibit Brownian movement in the regions of the smear where pockets of water are formed in the oil. This movement is an extremely valuable diagnostic character as the spores of other pathogens of honeybees usually remain fixed.

Stretch test is followed where the dead larval contents easily adhere to the tip of a pointed stick dipped into the larval extract and stretch like an elastic when lifted (Figure 4.2). Similarly, microscopic examination of infected larval scales stained with nigrosin show a mass of bacterial spores. Holst test essentially consists of placing the dried scales in warmed milk at 74°C. The milk curdles in about one minute and begins to clear within 15 minutes. This liquefaction of the milk is brought about by the action of an enzyme produced by spores of the *Paenibacillus*. Immunofluorescence,

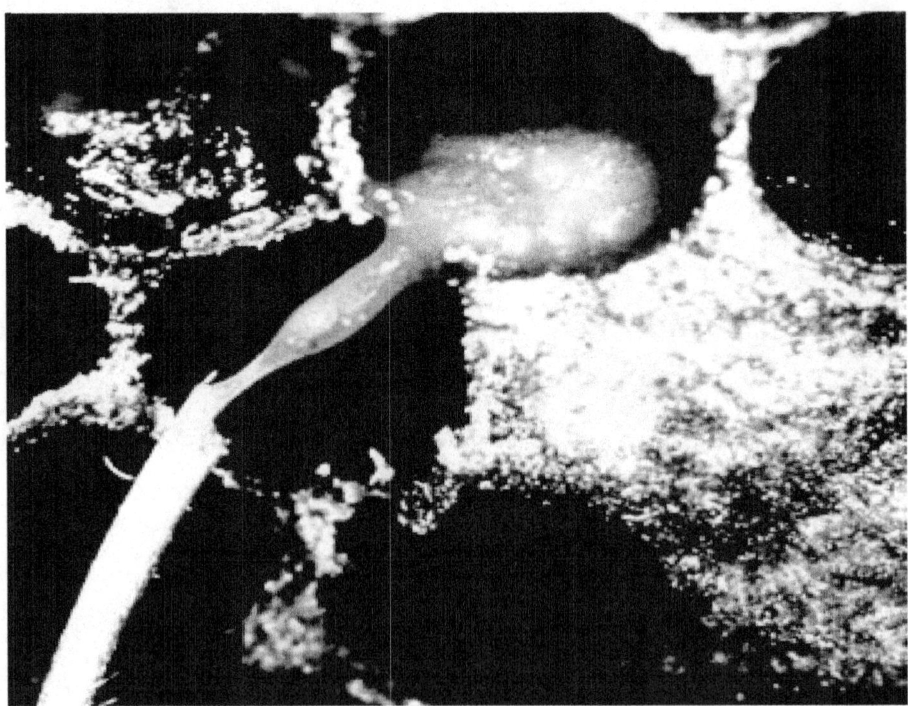

**Figure 4.2**   Diagnosis of American foul brood through stretch test

immunodiffusion (Peng and Peng, 1979) and use of monoclonal antibody in enzyme-linked immunosorbent assay (Olsen *et al.*, 1990) are the other diagnostic techniques readily used in the detection of American foul brood disease. Rabbits are injected with pure cultures of the pathogen and the active antiserum collected is stained with a fluorescent dye. This fluorescent antiserum is mixed with a bacterial smear on a slide and is examined with a fluorescent microscope where *P. larvae* appear as brightly fluorescent bodies on a dark background. Similarly, the spores can be observed at subclinical levels by the inoculation of honey in agar medium (Hansen, 1984). The fatty acids of the pathogen are also characterized by gas–liquid chromatography (Kaneda, 1969).

The disease spreads through contaminated equipment, drifting bees, spore-contaminated honey and pollen and robber bees. The spores of *P. larvae* transmit horizontally between the apiaries within the radius of one kilometre from the infected colonies (Lindstrom *et al.*, 2008). Though a few spores are quite sufficient to infect a day-old larva, millions of spores are required to infect an older larva. The queen larvae are more susceptible to disease than the larvae of worker and drone bees (Bailey and Ball, 1991).

## Management

***Manipulative methods*** Honeybee colonies could be placed in the areas rich in nectar and pollen flow during the active season. An efficient control of the disease could be possible by early detection of the pathogen and subsequent destruction of infected bee colonies. The queens of the diseased colonies may be replaced with mated healthy young queens, as they tend to be prolific and effectively maintain the hygienic colony conditions. As the bacterial spores can sometimes remain on honey and pollen, it is advisable to feed the colonies with spore-free honey and pollen (Morse and Hooper, 1985). Smirnov (1982) suggested not to use pathogen-infected combs for at least three years as the spores remain viable on the bee combs.

A beekeeping technique employed against American foul brood is artificial swarming or shook swarm method after honey flow seasons (Shimanuki and Knox, 1997). This technique involves transferring of adult bees to a disease-free hive followed by destroying diseased brood combs. Depleting adult bees supplied with comb foundation leads to a break in the survival of the pathogen in the absence of the brood. The depleted bees may eliminate most of the bacterial spores through defaecation. Later on the bees are no longer a source of infection (Hansen and Brødsgaard, 2003). This method is reported to be successful, but the disease may reoccur occasionally (Hornitzky and White, 2001).

***Sterilization of combs and equipment*** *Paenibacillus larvae* remains viable on the combs and equipment and may become a potential source of further infection. Sterilization of contaminated beekeeping equipment with a blowtorch is a common practice. However, sometimes this practice may not destroy all the spores successfully (Hansen and Rasmussen, 1990). The spores are extremely resistant to heat, freezing, common disinfectants and even to the germicidal action of honey (Hornitzky and Willis, 1983). It is possible that different strains of bacterium have varied tolerance to heat treatment (Calesnich and White, 1952). Gamma irradiation with cobalt-60 decontaminates the equipment for its use in the apiary (Baggio *et al.*, 2005). Higher doses of irradiation is known to inhibit the germination of spores on the larval scales and honey. The non-irradiated larval scales and honey have normal growth of the pathogen (Table 4.1).

**Table 4.1**    Viability of *Paenibacillus larvae* spores on exposure to heavy velocity electron beam radiation (Melathopoulos *et al.*, 2004)

| Radiation dose (KGy) | No. of drops inoculated | No. of drops with growth | No. of spores/ml $(\times10^9)\pm$(SE) | No. of viable spores/ml $(\times10^4$ cfu) |
| --- | --- | --- | --- | --- |
| 0 | 60 | 60 | 2.89$\pm$3.96 | 3.4$\pm$0.5 |
| 4.32 | 22 | 12 | 4.10 | 0.6 |
| 6.26 | 60 | 0 | 91.90$\pm$113.40 | 0 |
| 8.19 | 60 | 0 | 13.60$\pm$9.76 | 0 |
| 10.18 | 32 | 0 | 15.40 | 0 |
| 12.16 | 32 | 0 | 6.45$\pm$2.65 | 0 |

Fumigation of empty combs and equipment is probably the safest and most economical practice in management of bacterial diseases in honeybee colonies. The combs and equipment are fumigated with formaldehyde. Similarly, soaking of combs in mixtures of soap–formalin, alcohol–formalin and glycerol–formalin have also been advocated. The sterilized bee boxes may be treated with ethylene oxide (1 g/l) for two days in fumigation chambers after humidification (Shimanuki *et al.*, 1970). Placing hive parts and equipment in paraffin wax would deactivate the bacterial spores (Goodwin and Haine, 1998). The beeswax containing *P. larvae* on melting with steam in a wax processing plant was free from pathogen (Hansen and Rasmussen, 1990).

### *Breeding for disease tolerance*

**1. *Tolerance through food***    The bee pathogenic bacteria have developed tolerance against different varieties of antibiotics (Alippi, 1994) and breeding techniques have proved to be effective as alternative control measures. The disease tolerance in honeybees is known to be associated with their routine food. It is known that the queen larvae fed with least amount of pollen are most susceptible and worker larvae supplied with moderate amounts of pollen are intermediate but the drone larvae fed mostly with pollen are least susceptible to the disease (Rinderer and Rothenbuhler, 1969). Pollen contains a variety of microorganisms, which generally act as antagonist against the bacterial pathogens. These antagonist isolated from the midgut of the larvae and adult bees have proved to inhibit spore germination and growth of the *P. larvae* under *in vitro* conditions. It is likely that the inhibition occurs in the intestine of the larvae and one of the inhibitory components is fatty acid present in the royal jelly (Bilikova *et al.*, 2001). Most of the bacterial spores in the food suspension are filtered in the honey stomach of the bee by proventricular action and prevent the further infection of the larvae (Sturtevant and Revell, 1953).

**2. *Tolerance through hygienic behaviour*** The tolerance in honeybee colonies against American foul brood is determined by behavioural factors rather than physiological mechanism. The young larvae are known to be more susceptible to diseases than older larvae. Rothenbuhler (1964) compared the resistant lines of bees to a highly susceptible line and found that larvae from different genetic lines showed varied degrees of resistance. The hygienic behaviour is recognized as a classical example of Mendelian inheritance which is governed by genetic and environmental factors (Alcock, 1993). Rothenbuhler (1964) developed a two-locus model of inheritance for hygienic behaviour. The process of uncapping of a cell containing dead brood and removing the contents was thought to be dependent on two recessive genes at a homozygous condition. However, in a subsequent re-evaluation of Rothenbuhler's model, Moritz (1988) suggested a three-locus model, which shows the

involvement of many complex patterns of inheritance. The young resistant bees remove all the diseased brood irrespective of nectar availability, but bees older than a month remove the larvae only during a nectar flow season (Thompson, 1964). The variability in expression of this behaviour indicates that bees must carry genes for hygienic behaviour for colony level expression (Trump *et al.*, 1967).

The bee colonies representing many queen bee lines inoculated with spores of *P. larvae* responded in two ways. In the first group the colonies became severely infected. In the second group, the disease developed was low but later became disease-free. Such resistant colonies were bred and the resistance was inherited by daughter colonies. This indicates that the inheritance of resistance may be attributed either to physiological process in the larvae or the behaviour of adult bees. The cross breeding of hybrids of susceptible colonies with the resistant parent colonies had a quarter each of the offspring being either susceptible or resistant (Thompson and Rothenbuhler, 1957). A quarter colonies would uncap the cells but did not remove the larvae and the remaining quarter would remove the larvae only when the cells are uncapped. This behaviour is a potential tip for breeding resistant lines of bees against American foul brood. Another factor in the resistance mechanism is the ability of colonies to detect and remove diseased brood before sporulation of the pathogen. In a resistant colony, no spore formation was observed whereas in susceptible colonies, spore formation occurred on the bee larvae (Woodraw and Holst, 1942).

***Chemotherapy*** Antibiotics are known to suppress the development of the bacterial pathogens though they may not destroy spores successfully. Similarly, these are not used in the colonies during honey flow seasons as they contaminate honey and other hive products. Knox *et al.* (1976) demonstrated that, practice of shook swarm method followed by a single treatment of oxytetracyclin hydrochloride (Terramycin®) was effective in eliminating the disease. Sodium sulphathiazole (1.5 g/15 l) and oxytetracyclin hydrochloride (0.4 g in 5 l) suppress the disease when fed with strong sugar syrup. The fundamental action of sulphathiazole is undoubtedly bacteriostatic nature in many animals. Similarly, prophylactic use of drugs may help to eliminate an infection before spore formation (Ratnieks, 1992). According to Haseman (1948), penicillin and streptomycin are most effective in inhibiting the vegetative growth of *P. larvae* under laboratory conditions. Similarly, ampicillin readily inactivates the pathogen under *in vitro* conditions. Tylosin tartrate is an alternative to oxytetracyclin hydrochloride as its toxicity for larvae and adult bees is negligible (Peng *et al.*, 1996). A powdered sugar mixture with tylosin applied as a dust was quite effective in eliminating the disease. About 200 mg tylosin per 20 g of sugar powder is applied at weekly intervals for three weeks (Elzen *et al.*, 2002).

Azadirachtin is the active chemical component of the neem tree, *Azadirachta indica* A. Juss. and its formulated compounds (5 µg/ml) inhibit the cultures of *P. larvae* (Williams *et al.*, 1998). Similarly, natural vegetable oils (Alippi *et al.*, 1996) and fatty acids (Feldlaufer *et al.*, 1993) have been reported to suppress the growth of the pathogen under *in vitro* conditions. Among the five essential oils, the Andean thyme, *Acantholippia seriphioides* A. Gray is known to be the most effective against the pathogen. These oils have low toxicity on bees and mammals and also less harmful to the environment, which would be safer alternatives to the antibiotics.

# EUROPEAN FOUL BROOD

*Melissococcus plutonius* Corrig (formerly *Melissococcus pluton*) (Lactobacillales: Enterococcaceae), a non-spore-forming bacterium causes European foul brood. It is lanceolate in shape and occurs singly, in pairs and in chains. The bacterium is gram-positive and produces vegetative cells that measure about 1.0 $\mu$m in length and 0.5 to 0.7 $\mu$m in width. It is an infectious, contagious and stress-related disease primarily affecting the young larvae. This disease has been found to occur in all continents where *A. mellifera* colonies are placed. It is also reported in *A. cerana* from parts of central and southern states of India (Kshirsagar and Godbole, 1974). A few strains of *M. plutonius* have been isolated from larvae of *A. cerana* and one strain has been isolated from *A. laboriosa* (Bailey and Ball, 1991). The virulence of the pathogen is common in high brood rearing season and is more evident in bee colonies deficit in proteins.

## Symptoms

*Melissococcus plutonius* usually infects the larvae that are 2 to 3 days old. Larvae swallow the bacteria with spore-contaminated food. The bacteria multiply vigorously in the food mass of the larval midgut. They destroy the peritropic membrane and invade the intestinal epithelium (Shimanuki, 1990). The earliest symptoms of the disease are a slight yellow or grey discolouration and uneasy movement of the larvae. Most of the larvae die at the coiled stage, on the bottom of the cells. The dead larvae appear like a collapsed mass giving a melted appearance (Figure 4.3). This mass consists of a turbid fluid containing numerous bacteria. The transparent peritropic membrane is filled with bacteria in the form of opaque chalk white clumps (Bailey and Ball, 1991).

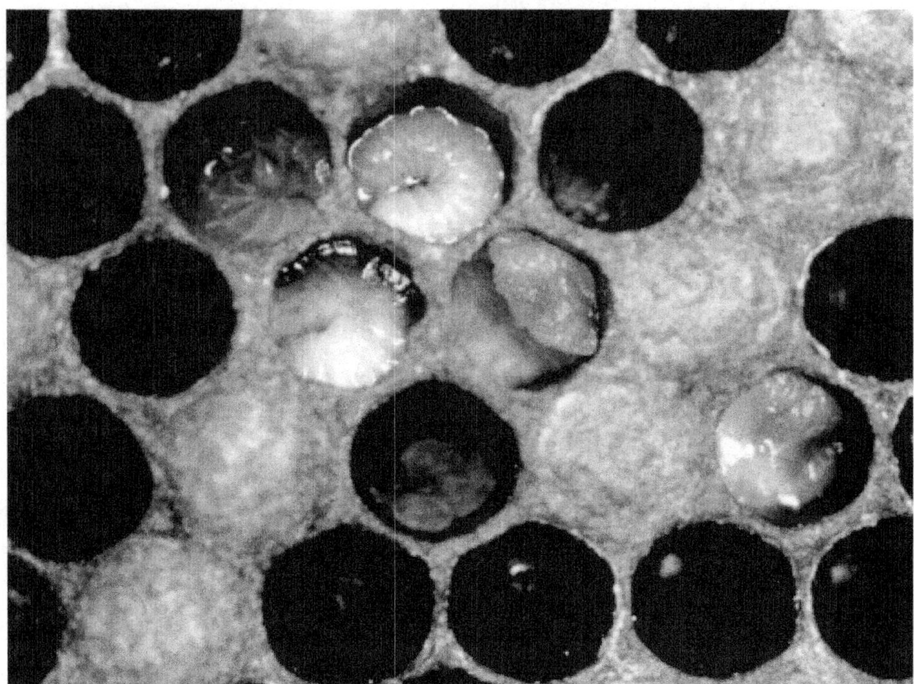

**Figure 4.3**   Honeybee larvae infected with European foul brood

The larvae undergo decaying, often giving off a foul odour. The dead larvae give off a sour odour and hence the disease is sometimes called as sour-brood. Dead and decaying larvae are usually not ropy as in American foul brood, however a slight ropiness is sometimes observed. An infected larva may spin cocoon with poorly developed silk glands but becomes flaccid and the tracheal system becomes quite visible. The diseased larva dries up into rubbery scales in the cell.

## Diagnosis

European foul brood could be diagnosed by exposing the smears of diseased larvae stained with carbol fuchsin under a microscope before appearance of secondary bacteria. The enzyme-linked immunosorbent assay (Pinnock and

Featherstone, 1984) and protocols of polymerase chain reaction have been developed for the detection of *M. plutonius* in honeybee larvae (Govan *et al.*, 1998). Polymerase chain reaction is also an alternative method for the detection of the pathogen in beehive products (McKee *et al.*, 2003).

The spores of the pathogen could remain viable and be able to cause infection for long periods. The contaminated combs and the faeces of surviving infected brood act as a source for further spread of the disease. Subsequently, the disease spreads through robber bees and by exchange of contaminated equipment between healthy and diseased colonies. Following the initial infection, several other bacterial pathogens, viz., *Paenibacillus alvei* Cheshire and Cheyne, *Enterococcus faecalis* Schleifer and Kilpper-Bälz and *E. faecium* Schleifer and Kilpper-Bälz (Lactobacillales: Enterococcaceae) would become secondary invaders. The death of infected larvae would also be accelerated by *Achromobacter eurydice* White (Eubacteriales: Achromobacteraceae) which is numerous in larvae infected with *M. plutonius*.

## Management

***Manipulative methods*** The severely infected colonies of European foul brood disease may be destroyed. The old queen may be replaced with young mated queen. Keeping bee colonies in the areas of good pollen and nectar flow is sometimes quite effective in elimination of the disease.

***Sterilization of combs and equipment*** Fumigation of empty combs and equipment is possibly the safest and most economical practice. Bee boxes are fumigated effectively by packing the bee combs in plastic bags filled with a mixture of ethylene oxide and carbon dioxide for a week. Similarly, fumigation of combs with acetic acid (80 per cent) inactivate the pathogen. Gamma irradiation with cobalt-60 would be effective in sterilizing the contaminated combs, honey, pollen and beekeeping equipment (Hornitzky, 1986).

***Breeding for disease tolerance***   Milne (1985) conducted studies on the hygienic behaviour against European foul brood in *A. mellifera* colonies and reported that the hygienic behaviour does not confer resistance, as there was no correlation between uncapping and removal in laboratory cages. However, the pathogen-infected larvae are invaded by secondary bacteria.

***Chemotherapy***   Antibiotics are known to be effective against *M. plutonius* with varied efficacy. However, occasionally, they may eliminate beneficial bacteria (Charbounneau *et al.*, 1992). Sodium sulphathiazole (1.5 g/15 l) suppress the disease on feeding with strong sugar syrup. Similarly, oxytetracycline hydrochloride (Terramycin®) may be fed or sprinkled with sugar syrup over the bee cluster in the hive during warm weather. Erythromycin, streptomycin and penicillin are other antibiotics found to be effective against European foul brood disease.

## PARA FOUL BROOD

The bacterium, *Bacillus paraalvei* Burnside (Bacillales: Bacillaceae) causes para foul brood disease in honeybees (Burnside, 1932). It is an aerobic, gram-positive, motile and spore-forming rod-shaped bacterium whose coccoid cells are tapered during the later stages. It is 2.0–5.0 $\mu$m long and 0.5–0.8 $\mu$m wide and its spores are 0.8 × 1.8–2.2 $\mu$m in size. The spores are generally clumped and show stacked appearance. Typical cultures of *B. paraalvei* spread vigorously on nutrient agar with motile colonies. The worker, queen and drone larvae and sometimes pupae are affected by para foul brood disease.

## Symptoms

The larvae infected with para foul brood would become slightly less plumpy and change in colour from glistening white to a dull white. The cell capping becomes dark, sunken and greasy. The infected brood produces a sour odour. A large number of larvae are coiled or irregularly twisted in the cells, although many larvae die in an extended position. Later, the larvae turn reddish brown and form dark coloured scales. Since the epidemiology of para foul brood is almost similar to that of European foul brood, similar management practices would be effective against para foul brood also.

## SEPTICAEMIA

Septicaemia is associated with adult bees caused by a bacterium, *Pseudomonas apiseptica* Burnside (Pseudomonadales: Pseudomonadaceae). It is a gram-negative pathogen occurs singly, in pairs or in short chains (Landerkin and Katznelson, 1959). Septicaemia is most prevalent in the bee colonies placed near moist soil. The pathogen infects queens, drones and worker bees through the first pair of thoracic spiracles (Shimanuki, 1990).

## Symptoms

The symptoms of septicaemia are a change in the colour of the haemolymph from apple brown to chalky white followed by rapid regeneration of muscles in adult bees. The diseased bees become weak and are unable to fly. Severe infection causes the haemolymph to become turbid and milky. This results in disintegration of bees with a characteristic imbalance in the movement of legs, wings and antennae. Dead or dying bees are known to emit a putrid odour.

The transmission of the pathogen seems to be possible through contaminated water. There are several species of bacteria that cause septicaemia when placed on the haemolymph of bees, but only *P. apiseptica* cause small spherical cells. *Serratia marcescens* Bizio and *Hafnia alvei* Moller (Enterobacteriales: Enterobacteriaceae) have also been reported to cause septicaemia and are transmitted through the mite, *V. destructor* (Strick and Madel, 1988).

Since moisture favours the spread of infection, selecting well-drained apiary sites and exposing colonies to sunlight for at least a part of the day would minimize the disease. Streptomycin has been used to control septicaemia in Switzerland, but the development of resistance to these antibiotics in some strains of the pathogen has limited its use (Shimanuki and Knox, 1991).

## SPIROPLASMAS AND MYCOPLASMAS

*Spiroplasma* spp.(Entomoplasmatales: Spiroplasmataceae) cause spiroplasmosis in adult honeybees. These bacteria are helical, motile, prokaryotes found in the haemolymph of adult bees. They are tiny, coiled and sometimes branched with a filament of 0.7 to 1.2 $\mu$m in diameter. The grown bacteria ranges in length from 2 to 10 $\mu$m. *Spiroplasma* infects worker and queen bees on feeding or inoculation. The May disease is known to be associated with *Spiroplasma apis* Mouches and the infected bees have swollen abdomen and show agitated movement. These bees crawl down from the hives in large groups and die abundantly during the month of May (Mouches *et al.*, 1984) and thus the name. The pathogen could be seen in the haemolymph with dark-field microscope. Mycoplasmas are rounded prokaryotes. These particles generally resemble mycoplasma of the plants.

## OTHER BACTERIAL PATHOGENS

*Bacillus pulvifaciens* n. sp. (Bacillales: Bacillaceae) is known to cause the disease rarely in honeybees. The infected larvae are light brown to yellow in colour. Similarly, *Pseudomonas fluorescens* Migula (Pseudomonadales: Pseudomonadaceae) was recorded on honeybees (Katznelson, 1950). Bacteria such as *Bacillus laterosporus* (Bacillales: Bacillaceae) and *Chromobacter eurydice* are commonly associated with honeybees. *Paenibacillus larvae* subsp. *pulvifaciens* n.sp. (Bacillales: Paenibacillaceae) is suspected to cause powdery scale disease in honeybees. The powdery scale is light brown to yellow and extends from the base to the top of the cell and crumbles into dust.

# 5

# Fungal Diseases

Insects are known to suffer from many species of fungal pathogens but most of them are species-specific. Generally moderately warm conditions with high humidity favour the growth of fungi. They infecst brood, adult bees and combs containing stored products in honeybee colonies. The most common fungal diseases of honeybees are chalk brood and stone brood.

## CHALK BROOD

The fungus, *Ascosphaera apis* (Maassen ex Claussen) Olive and Spiltoir (Ascosphaerales: Ascosphaeraceae), causes the chalk brood disease. It infects larval and pre-pupal stages of the bee brood. It is generally stress-related but the damp conditions with poor hive ventilation favour the intensity of disease. The chalk brood is frequently found to occur in spring and early summer by causing severe damage to bee colonies. It occurs widely in the temperate regions in *A. mellifera* colonies. It has also been recorded infecting the pupae of *A. cerana* (Gilliam *et al.*, 1993).

This heterothallic fungus produces spore cysts through sexual reproduction from both + (plus) and – (minus) strains that are grown in close proximity. The spore cysts are olive green to brown in colour and globose in shape (Figure 5.1). They measure $47 \times 140\,\mu$m in size and bear many spore balls of 11–17 $\mu$m. The spores are $3.0$–$4.0 \times 1.4$–$2.0$ $\mu$m in diameter (Bailey and Ball, 1991).

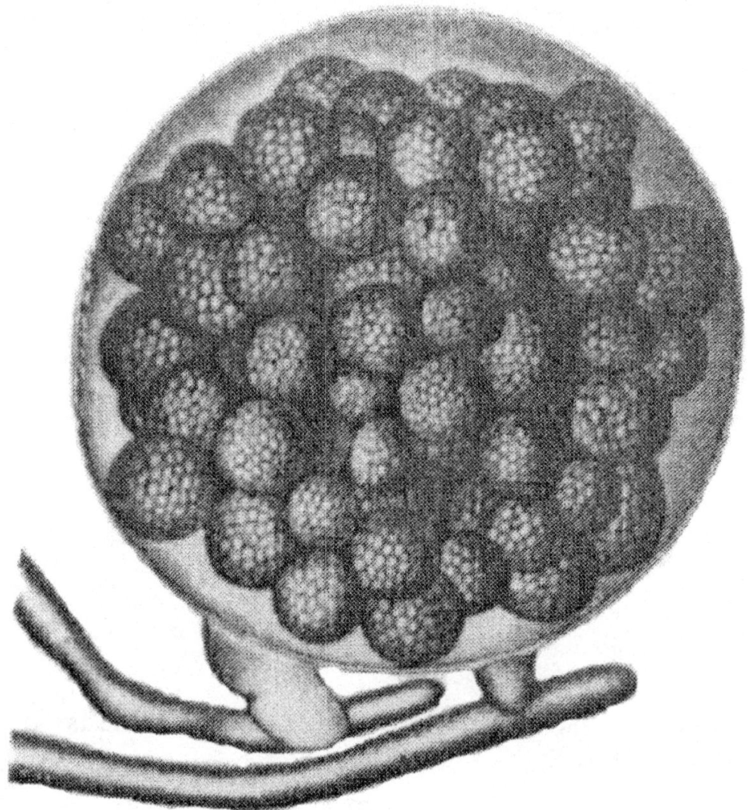

**Figure 5.1**    Spore cyst of *Ascosphaera apis* containing spore balls

## Symptoms

Chalk brood disease can be easily identified by its gross symptoms. The fungus infects younger larvae and pre-pupae usually located in outer fringes. The spores are ingested by

larvae along with food and they germinate in the lumen of the gut. The infected larva dies after cell capping and turns white followed by grey and black colour on formation of the fruiting bodies. It is overgrown by fluffy mycelia and swells. The infected larva dries into hard, shrunken white chalk mummies and hence the name chalk brood, when only one thallic strain (+ or –) is present. When both the thalli (+ and –) are present, the mummies formed are either mottled or completely black-coloured (Gilliam *et al.*, 1978). The black mummies are the result of fruiting bodies and the white mummies are due to the growth of fungal mycelium (Figure 5.2). In the hives with large number of infected larvae, mummies can be found at the hive entrance or on the bottom board. Mummies can sometimes be removed from brood cells by tapping the comb against a

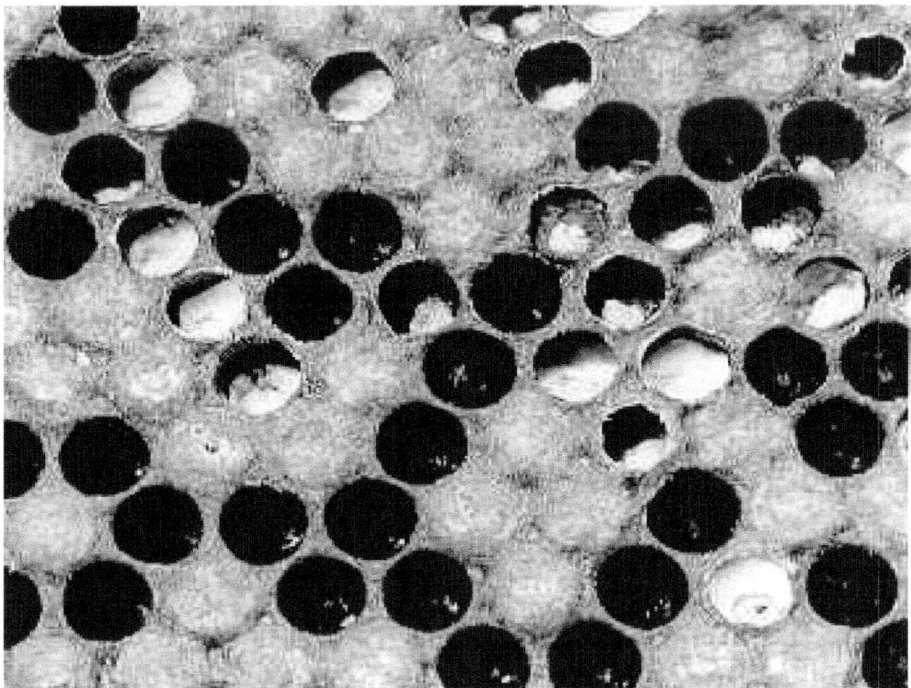

**Figure 5.2**   Hard chalklike mummified larvae of honeybee, *Apis mellifera*

solid surface. Thus easy removal of larval remains also differentiate chalk brood from other brood diseases.

## Diagnosis

The fungal pathogen, *A. apis* can be diagnosed by observing stained mummies containing spore cysts under the microscope. A rapid detection and identification of the pathogen could also be possible by a polymerase chain reaction technique using specific primers.

The fungal spores may be present in honey and pollen stored in the infected colonies. They remain viable on the combs and eventually germinate during favourable conditions. The bees remove the diseased larvae in their routine cleaning process. During this period the spores may enter into healthy larvae. The transmission of disease occurs through contaminated pollen and water sources in apiary conditions (Koenig *et al.*, 1987). It could also be transmitted by drifting bees, wind, water, birds and the beekeepers (Thorstensen, 1976).

## Management

***Manipulative methods***   Strengthening of weak colonies by uniting adult bees and brood combs and providing sufficient ventilation would prevent the chalk brood. Periodic renewal of old combs is most effective, as the pathogen could remain infective on these combs for longer periods. Nelson and Gochnauer (1982) found reduced chalk brood infection on new combs. The hives could be placed in well-drained areas away from larger water bodies.

***Sterilization of combs and equipment***   Ethylene oxide and methyl bromide have been used to fumigate hive equipment and combs infected with chalk brood (Gochnauer and Margetts, 1980). Gamma irradiation with cobalt-60 sterilizes hive parts and bee products free of viable chalk brood spores (Katznelson and Robb, 1962). Mummies of *A. apis* that had been buried less than 8 cm deep in pollen did not produce growth following exposure to high velocity electron beam radiation (Table 5.1). Similarly Anderson *et al.* (1997) recommended the sterilization of honey in a water bath at 65°C for 8 hours or 70°C for 2 hours.

**Table 5.1** Viability of *Ascosphaera apis* in dried pollen at different depths exposed to high velocity electron beam radiation (Melathopoulos *et al.*, 2004)

| Depth (cm) | Radiation dose (kGy) | No.of Mummies in sample | Samples showing viable mycelia |
|---|---|---|---|
| Not irradiated | | 5 | 5 |
| 2 | 10.5 | 3 | 0 |
| 4 | 11.0 | 3 | 0 |
| 6 | 11.5 | 3 | 0 |
| 8 | 6.5 | 3 | 0 |
| 10 | 2 | 3 | 3 |
| 12 | 0 | 1 | 1 |
| 14 | 0 | 3 | 3 |
| 16 | 0 | 3 | 3 |
| 18 | 0 | 3 | 3 |
| 20 | 0 | 2 | 2 |

***Breeding for disease tolerance*** Honeybees have developed several immune mechanisms against fungal diseases. They have cuticles impregnated with waxes and unsaturated fatty acids. The chitinous lining of gut acts as a protective barrier against microbes which are engulfed and digested in the phagolysosome through hydrolytic enzymes. The larvae of hygienic colonies filter the fungal spores at a faster rate through proventriculus. Bees may also add antagonistic moulds and bacteria (*Bacillus* spp.) into the pollen, which inhibit the growth of chalk brood pathogen (Gilliam, 1990).

Hygienic behaviour is known to be the primary mechanism of resistance against chalk brood although it involves many factors (Spivak and Gilliam, 1993). The hygienic colonies remove the larvae infected with chalk brood within a day. Similarly physiological resistance may also exist against the disease (Spivak and Gilliam, 1993). Gilliam *et al.* (1983) found that colonies with good hygienic behaviour are less likely to become infected on inoculation of sugar syrup contaminated by fungal spores. Therefore, selection of hygienic colonies holds the most promising method for the control of chalk brood. Spivak (1996) demonstrated that there were no chalk brood mummies in selected parents and daughter colonies mated naturally with unselected drones. Oldroyd (1996) also developed a stock of bees tolerant to chalk brood disease in Australia.

**Chemotherapy**   Trichloroisocyanouric acid (TCA) dissolved in water is effective in control of chalk brood disease (Tanaka *et al.*, 1984). Certain alkyl amines stimulate the removal of chalk brood cadavers by worker bees and inhibit the growth of the fungus under *in vitro* conditions (Herbert *et al.*, 1986). Feeding an essential oil of *Satureia montana* Linn. (Labiaceae) mixed with sugar syrup reduces the intensity of the disease (Colin *et al.*, 1989). Similarly, the cinnamon oil and thymol inhibit the growth of the pathogen within a week (Calderone *et al.*, 1994). Bee colonies treated with neem extract containing low concentration of azadirachtin ($3\ \mu g/1$ ml of syrup) produced smaller number of chalk brood mummies (Liu, 1995).

## STONE BROOD

Stone brood disease is generally caused by the fungus, *Aspergillus flavus* Link (Eurotiales:Trichocomaceae) and occasionally by *Aspergillus fumigatus* Fresenius (Eurotiales: Trichocomaceae). These common soil inhabitants are pathogens of many insects, birds and mammals. It is more prevalent in beehives with poor ventilation under damp conditions.

## Symptoms

The spores of the pathogen germinate on the cuticle of larva and the mycelium occasionally penetrates the sub-cuticular tissues. It produces local aerial hyphae and conidiospores (Figure 5.3). The spores are found abundant near the head

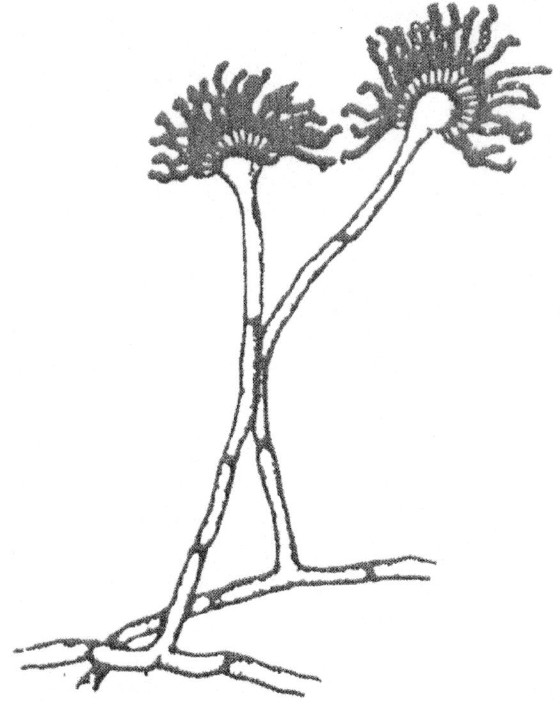

**Figure 5.3**  Conidial heads of *Aspergillus flavus* as developed on honeybees

of the infected larvae and pupae and form green stonelike solid mummies. The larva becomes hardened and quite difficult to crush after its death, hence the name stone brood. The aflatoxins of the fungus also act on the central nervous system and endocrine system of the bees (Burnside, 1932).

The colony manipulative methods used for the control of chalk brood disease may also be effective against the stone brood in honeybee colonies.

## OTHER FUNGAL PATHOGENS

Some species of yeasts have been isolated from the intestine of infected bees. One such pathogenic yeast is *Torulopsis* spp. (Mucorales: Mucoraceae), closely related to *Torulopsis candida* Lodder (Skou and Holm, 1980). The pollen mould, *Bettsia alvei* Skou (Ascosphaerales: Arthrodermataceae) is found in bee colonies on the pollen stored combs especially in temperate regions. Both the yeasts and moulds are of minor importance as fungal pathogens of honeybees.

# 6

# Protozoan Diseases

Protozoans are a unicellular, diverse group of eukaryotic organisms and about 500 species are known to be pathogens of insects. Many protozoans debilitate the host without showing visible symptoms, while some species are extremely pathogenic by causing stunted growth and premature death of the host (Tanada and Kaya, 1993). They are either parasitic or symbiotic on honeybees and cause greater losses to the beekeeping industry throughout the world. Microsporidia are obligately intracellular organisms that are common parasites of insects and few invertebrates. The microsporidian and protozoan diseases of honeybees are nosemosis and amoeba disease respectively.

## NOSEMOSIS

Nosemosis is one of the most widespread adult honeybee diseases caused by the microsporidian parasite, *Nosema apis* Zander (Microspora: Nosematidae). It is ubiquitous and occurs in large numbers in bee colonies during brood-rearing season. It is distributed worldwide and has been reported from many parts of India on *A. cerana* colonies (Kshirsagar and Godbole, 1974).

## Symptoms

Bees infected by *Nosema apis* do not show any external disease symptoms. However, the symptoms expressed are not only due to nosemosis but are also attributed to other diseases of bees. *N. apis* is an obligate parasite which develops in the gut tissues of adult bees and has been known to shorten the lifespan of honeybees. The infected bee colonies show restlessness and are unable to fly but drop a loose excreta on the combs and hive parts. The affected crawler bees become abundant in front of the hive. The hindwings of bees may get unlocked from the forewings and held at unusual angles. The infected nurse bees do not produce sufficient royal jelly due to deterioration of food glands. The hypopharyngeal glands of newly emerged adult bees with the spores of the pathogen fail to develop completely and eventually undergo atrophy. The midgut of bees become swollen due to accumulation of spores. The intestine of the heavily infected bees are likely to appear dull greyish white with or without circular constrictions. As the infection advances, formation of the striated border and the peritrophic membrane in the intestine will be hindered. Nosemosis causes degeneration of oocytes in the ovary of the queen. The number of eggs laid by the queen is reduced and occasionally the queen dies on severe infection.

## Diagnosis

Nosemosis can be diagnosed by microscopic examination of the ventriculus of the infected bees. The ventriculus, which is normally brown in colour, becomes white and fragile on infection. Giemsa-stained (10% 0.02 M phosphate buffer)

air-dried, ethanol-fixed smears of infected tissues show spores with thick unstained walls without visible nuclei. The spores can also be detected in the faecal matter.

Transmission of the spores is possible by ingestion of contaminated food and water. Similarly, robbing, uniting of bee colonies and exchange of hive parts between healthy and diseased colonies would also spread the disease.

## Biology

*Nosema apis* multiplies exclusively in the ventriculus of the midgut epithelial cells of adult honeybees (Morgenthaler, 1963). *Nosema* spores are dark oval structures and are about 4 to 6 $\mu$m long and 2 to 4 $\mu$m wide (Figure 6.1). The spores enter the lumen of the midgut of bees through the mouth along with contaminated food or water. The spore germinates

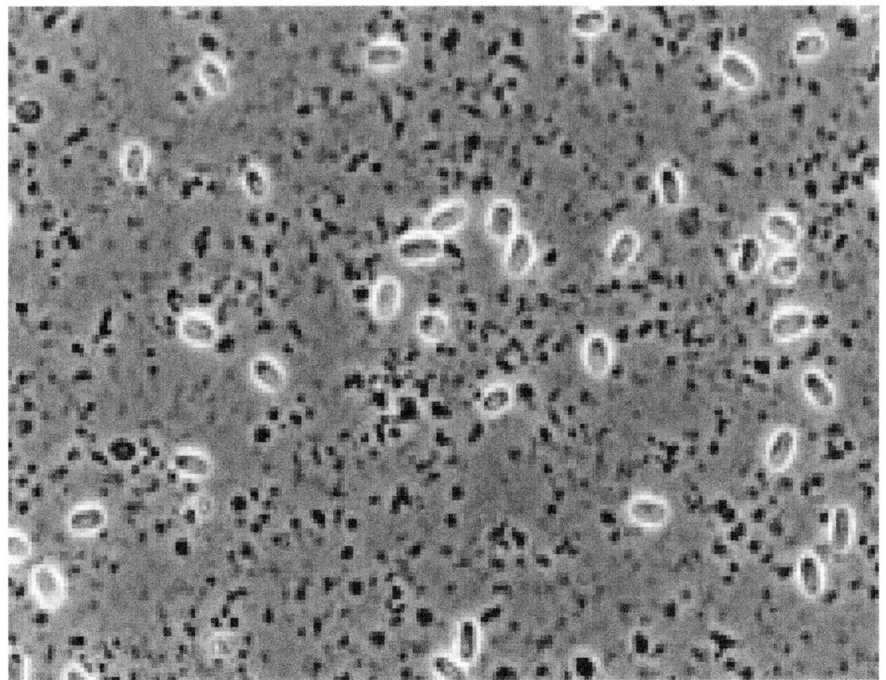

**Figure 6.1**   Spores of the microsporidian, *Nosema apis*

by fusion of the nuclei and the stage of the parasite is called as planont, which creeps slowly on the epithelial cells. Many planonts may enter a single host cell. After certain nuclear changes, the planont becomes the meront. The meront increases in size and multiplies by fission. The meronts become sporonts and by the process of sporogony, spores are formed from the sporonts. The spores are shed into the bee's gut cavity, eliminated into the rectum and are finally excreted through faeces. Workers, queens and drones of all ages are susceptible to the disease. The entire life cycle of the parasite is completed in three to four days.

## Management

***Manipulative methods***  Maintenance of strong bee colonies with a prolific queen and sufficient food stores can prevent nosemosis. Extraction of honey and subsequent feeding of sugar syrup with lower ash content will reduce the risk of dysentery. Old combs which are constant sources of pathogens may be replaced with new combs. Protein deficiency possibly increases the level of nosema infection and hence providing high protein-rich pollen would overcome the protein deficiency as well as occurrence of the disease.

***Sterilization of combs and equipment***  Combs without honey or pollen may be disinfected with formalin. Fumes of acetic acid (60 per cent) would inactivate the *Nosema* spores. The acetic acid is interspersed with absorbent material between the piles of hive containing combs. Similarly, exposing the hive equipment to a temperature of at least 60°C for 15 minutes and combs to 49°C and 50 per cent relative humidity for a day reduces nosemosis without damaging the combs (de Ruiter and Steen, 1989).

***Chemotherapy***  The antibiotic, fumagillin, is highly effective against *N. apis* (Woyke, 1984a). It is derived from the fungus, *Aspergillus fumigatus,* and available under the trade name, Fumidil B. It suppresses *Nosema* infection when fed to bees at concentrations of 0.5 to 3 mg/100 ml sugar syrup (Bailey and Ball, 1991). However, it may not be effective against dormant *Nosema* spores. Feeding Fumidil B to the bees during spring may not have a long-term effect than autumn as the spores deposited on combs during winter are unaffected. Similarly, mercurial compounds including sodium ethylmercuric thiosalicylate (Merthiolate) are also effective against the parasites.

Protofil is a product of essential oils rich in cyclic and aliphatic hydrocarbons. It is extracted from the plants such as *Teraxacum officinale* Wigg. (Dandelion), *Thymus vulgaris* Linn. (Savony), *Achillea millefolium* Linn. (Milfoil) and *Ocimum basilicum* Linn. (Basil). It stimulates the secretion of digestive enzymes in honeybees and inhibits the growth of *N. apis* on feeding (Chioveanu *et al.,* 2004). Other drugs recommended are gramicidin, sulphaquinoxaline, anisomycin, phenol urotropin, B-naphthol, salol and colloidal sulphur. The potential of some natural compounds such as thymol, vetiver essential oil, lysozyme, resveratrol for the control of nosema infection in honeybees was evaluated. The bees fed with thymol and resveratrol candies had significantly lower rates of infection (Maistrello *et al.,* 2008).

## *Nosema ceranae*

*Nosema ceranae* n. sp. (Microspora: Nosematidae) was initially reported to be a pathogen of *A. cerana* and subsequently recorded on *A. mellifera.* It causes loss to beekeeping in a few European countries (Germany, Poland, Italy, Spain, Austria) and in New Zealand. *N. ceranae* and *N. apis* have similar morphometric characteristics under an optical microscope (Fries *et al.,* 1996). Huang *et al.* (2007)

found *N. ceranae* in the colonies of *A. mellifera* from Taiwan. This pathogen is known to be linked with colony collapse disorder primarily in the USA. The disease normally afflicts adult bees where the population of the colonies will decline and the bees can die within a week after exposure to the pathogen. The most effective control measure of *N. ceranae* is the application of the antibiotic, fumagillin (Williams *et al.*, 2008).

## AMOEBA DISEASE

Amoeba disease is caused by the protozoan, *Malpighamoeba mellificae* Prell (Amoebida: Entamoebidae). It is widely distributed in temperate regions but found to be absent from tropical and subtropical regions of the world including India. The *M. mellificae* cysts ingested by the adult bees germinate in the intestine of bees possibly at the posterior end of the ventriculus where solid food particles are accumulated. The cysts are 5 to 8 $\mu$m in size. It is known to be associated more frequently with nosemosis in bee colonies.

### Symptoms

The general symptoms of the amoeba disease are the gradual decline in the bee population. *M. mellificae* infects the lumen of the malpighian tubules of adult bees. During severe infection, the lumen of most of the tubules are packed with ovoid transparent cysts (Figure 6.2). These tubules are slightly distended, glassy in appearance and easily broken. The infection causes the epithelium of malpighian tubules to undergo atrophy (Liu, 1985).

### Diagnosis

A laboratory method developed for detection of *M. mellificae* consists of examination of suspected bees for the presence of cysts. The abdominal suspension of the suspected bees is

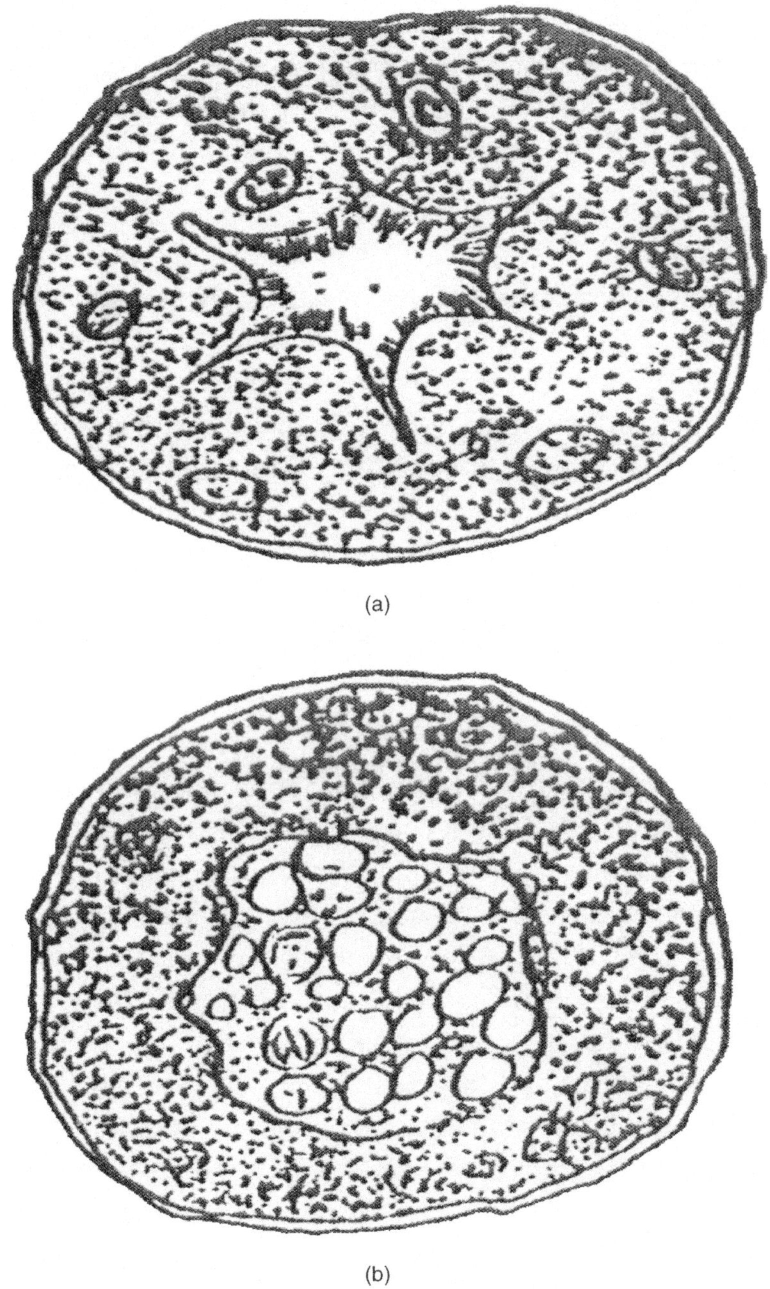

(a)

(b)

**Figure 6.2**   Cross sections of malpighian tubules (a) Healthy tubules (b)Tubules containing cysts of *Malpighamoeba mellificae*

prepared by maceration in distilled water and the filtered suspension is examined under the microscope where cysts can be seen clearly. Transmission of the parasite takes place through ingestion of cysts by bees, through faecal matter rich with the pathogen on the comb and also by robbing bees.

## Management

Amoeba-infected colonies are shifted to places rich in pollen and nectar. Old combs are a constant source of disease and renewal of such combs is recommended. Good management practices by wintering strong colonies with more quantity of well-ripened honey stores and maintaining hygienic conditions in the colonies prevents the prevalence of the disease. Some chemical compounds such as phenyl salicylate, quinosol, fumagillin, furazolodone and dichloroxyquinaldine are known to be effective against the amoeba disease.

## GREGARINES

Gregarines (Eugregarinida: Septatina) are occasionally associated with honeybees. Stejskal (1973) recorded four species of gregarines on bees from Venezuela. They are *Monoica apis*, *Apigregarina stammeri*, *Acuta rousseaui* and *Leidyana apis*. The sporozoites of the gregarines gain entry into the epithelial cells lining the gut. Immature stages, the cephalonts, are about $16 \times 44$ $\mu$m in size. They are oval with distinct body segments where the posterior segment is larger. Mature stages, the sporonts, are about $35 \times 85$ $\mu$m in size with reduced interior segment. Honeybees ingest the spores while cleaning up the faecal matter deposited in the hive. On germination, the cephalonts are attached to the bee's ventriculus. The colonies with infected bees are known to function abnormally resulting in dwindling. *Leidyana* spp. (Leidyanidae) encyst in the lumen of the midgut of adult

bees but are not very specific to honeybees (Wallace, 1966). Gregarines can be detected under low-power compound microscope.

## FLAGELLATES

The flagellates, *Crithidia* (*Leptomonas*) spp. (Trypanosomatida), infect the intestinal wall of honeybees (Lom, 1964). They move freely in the gut lumen or are attached to the epithelium of the hindgut or rectum of the adult bees. They vary in size from 5 to 30 $\mu$m and some appear as pearlike bodies with flagella. These are commonly found in Europe and Australia, where they are named as *Crithidia mellificae* and are not very specific to honeybees.

# NON-INFECTIOUS DISORDERS

Brood and adult stages of honeybees are occasionally affected by many non-infectious disorders other than pathogens or pests. These may be due to digestive ailments, poor colony management, fluctuation in the maintenance of constant hive temperature and poisoning of pollen and nectar while spraying of insecticides on host plants.

## BEE POISONING

The widespread use of synthetic organic pesticides has caused an unprecedented entry of chemicals into the environment, which threatens the integrity of the bee–flower mutualistic system. The pesticides are hazardous and cause large-scale mortality of bee pollinators. Bee poisoning occurs on field application of pesticides during blooming stage of bee flora where foraging bees collect and bring the poisoned nectar or pollen to the nest (Tasei *et al.*, 1987). Some pesticides are known to be persistent on the pollen and nectar and thus cause the large-scale mortality of bee foragers (Figure 7.1).

Pesticide-contaminated pollen and nectar cause bees to cease feeding and thereby reducing consumption and collection of nectar (Fiedler, 1987).

**Figure 7.1**   Dead honeybee with broken wings due to pesticide poisoning on the sunflower

The nectar and pollen from *Tilia* spp. are sometimes toxic for their high content of the trisaccharide, melezitose, although bees are able to digest the sugar. Similarly, honeydew from conifers and the nectar of *Rhododendron* spp. containing andromedotoxin are also poisonous to bees. Bees are killed by effluents containing arsenics and disposed by industries into the environment (Svoboda, 1962). Sulphur dioxide contamination may interfere with the olfactory system of honeybees (Bromenshenk, 1979).

Typical symptoms of bee poisoning are stupefaction, paralysis and disorganized behaviour in bees. The death of large number of adult bees near the hives with protruded tongue can be seen. Brood may be killed on being fed by contaminated nectar and pollen. Nation *et al.*, (1986) observed reduction in brood-rearing activity due to smaller doses of pesticidal application. Pesticides can also cause physiological imbalance and reduce longevity of bees (Smirle *et al.*, 1984). Depending on the type of pesticides used under field conditions, foraging bees may die on the return flight to the hive.

## DYSENTERY

Dysentery is a kind of defaecation by adult bees as they cannot ingest the crystals while feeding on coarsely granulated honey. It is also caused due to accumulation of water contaminated with toxins in the rectum. The most striking sources of toxins are acid-inverted sucrose and acid-hydrolysed starch that are marketed as sugars for feeding the bees. Hydrolysed sucrose though nutritionally better may cause dysentery in honeybees (Bailey, 1966). Dysentery is also associated with nosemosis where the affected colonies become weak and may even succumb to viral infection.

## NEGLECTED BROOD

The nurse bees are supposed to feed sufficient food to the larvae and maintain a constant temperature in the colony. However, sometimes, there may be sudden loss of adult bees in the colony and the left-over population may be insufficient to feed the larvae and regulate constant nest temperature. During these situations, the brood may die on a large scale due to insufficient population of nurse bees.

## CHILLED BROOD

Chilled brood is often observed during brood-rearing season, when the adult bees are unable to cover the entire brood and maintain a constant brood temperature. During colony manipulations, many empty frames fixed with comb foundation sheets are introduced in the brood chamber. The queen lays eggs on all the frames supplied but the adult population may not be sufficient to maintain the brood temperature. This leads to the outer frames of brood becoming chilled. Chilled larvae and pupae become yellowish tinged with black colour on the margins of the comb.

## STARVED BROOD

The larvae and pupae in honeybee colonies are often removed or eaten by worker bees occasionally due to floral dearth. This causes decline of adult bee population which leads to insufficient number of worker bees to feed the brood as a result brood starvation occurs. Sometimes newly emerged adult bees become starved and die with their tongues protruded.

## OVERHEATING

Honeybee brood may be overheated due to improper maintenance of hive temperature. This is possible in the summer season on bee colonies that are placed in the field under bright sunshine. Similarly, lack of sufficient ventilation also raises the temperature in the hives. This may occur due to declined nectar and water foragers. The overheated larvae are found hanging out of their cells. They turn brown to black in colour with a watery consistency. The brood cells may appear dry and result in death of the larva.

## HEREDITARY FAULTS

Eggs that fail to hatch into larvae and pupae in honeybees die with no apparent infection as they may be suffering from hereditary faults. Mackensen (1951) found irregularly developing brood in the combs and opined that it was due to disappearance of diploid drone brood. All the fertilized eggs were laid in worker cells, but half of them were believed to be homozygous at a sex determining locus of a chromosome and therefore such males were believed to be not viable.

## COLONY COLLAPSE DISORDER

Colony collapse disorder is a phenomenon in which worker bees from a colony abruptly disappear. It was originally found in *A. mellifera* in North America during 2006. The cause of the disorder is assumed to be associated with stress, pests, bee poisoning, etc. The symptoms include complete absence of adult bee population in colonies with a few dead bees in and around the colonies and presence of queen with capped brood cells. The colonies contain sufficient quantity of brood, pollen and honey. It is known to occur in India, Brazil and parts of Europe.

## MANAGEMENT

The most important part of preventing bee poisoning is the thorough understanding and cooperation between beekeepers and the farmers. Pesticides that are hazardous to bees should not be sprayed during blooming periods on crops. Pesticides posing lesser harm to bees could be sprayed on cross-pollinated crops preferably during dusk after the foraging activity of honeybees is reduced. During bee poisoning, colonies may be moved away from the area of pesticide application and fed with sugar syrup as a temporary measure. Frames that have been contained with pesticide-contaminated nectar or pollen may be replaced with new combs.

Bee colonies are placed in shade with sufficient ventilation to prevent the problem of overheating. The colonies may be provided with access to water and fed with sugar syrup. Removal of honey in the colonies, before granulation, prevents dysentery. Colonies may be maintained populous to minimize the problems of neglected brood, chilled brood, starved brood and over-heating. The colonies with collapse disorder may not be combined with healthy colonies. The sugar syrup can be fed to such colonies with an antibiotic, fumagillin as an alternative to reduce the intensity of disorder.

# PARASITIC MITES

Mites are the most diverse group of arachnids, and among arthropods they are second only to the insects in terms of number of species. Beehives provide an ideal habitat for many groups of organisms, of which mites are the most common inhabitants. They have established a greater extent of association with honeybees and threaten both domesticated and feral bee colonies. About 100 species of mites are associated with different species of honeybees (Eickwort, 1988). Three common suborders of mites associated with bees are Astigmata, Prostigmata and Mesostigmata. Based on their association with bees they are classified into four groups namely scavengers, predators on scavengers, phoretic and parasitic mites (Eickwort, 1997). The scavengers and predators on scavengers are of minor importance in bee colonies. Only the phoretic and parasitic mites show a greater association with honeybees and hence they are considered important in the studies on honeybees. The common phoretic and parasitic mites of honeybee species are presented in Table 8.1.

**Table 8.1**　Common phoretic and parasitic mites associated with honeybee species

| Mite species | Honeybee species | Reference |
| --- | --- | --- |
| **Ameroseiidae** | | |
| *Neocypholaelaps indica* Evans | *Apis mellifera* | Eickwort (1988) |
| | *Apis cerana* | Ramanan and Ghai (1984) |
| *Afrocypholaelaps africana* Evans | *Apis mellifera* | Grobov (1974) |
| **Tarsonemidae** | | |
| *Acarapis woodi* Rennie | *Apis mellifera* | Bennet (1980) |
| | *Apis cerana* | Lindquist (1968) |
| *Acarapis externus* Morgenthaler | *Apis mellifera* | Ibay (1989) |
| *Acarapis dorsalis* Morgenthaler | *Apis mellifera* | Bennett (1980) |
| *Pseudacarapis indoapis* Lindquist | *Apis cerana* | Sumangala and Haq (2002) |
| **Varroidae** | | |
| *Varroa destructor* Anderson and Trueman | *Apis mellifera* | Anderson and Trueman (2000) |
| | *Apis cerana* | |
| *Varroa jacobsoni* Oudemans | *Apis cerana* | Nagaraja and Reddy (1999) |
| | *Apis koschevnikovi* | Delfinado-Baker *et al.* (1989) |
| | *Apis nuluensis* | de Guzman *et al.* (1996) |
| *Varroa underwoodi* Delfinado-Baker and Aggarwal | *Apis cerana* | Delfinado-Baker and Aggarwal (19 |
| | *Apis nigrocincta* | Anderson *et al.* (1997) |
| | *Apis nuluensis* | de Guzman *et al.* (1996) |

| *Euvarroa sinhai* Delfinado and Baker | *Apis florea* | Kapil and Aggarwal (1987) |
| | *Apis andreniformis* | Delfinado-Baker *et al.* (1989) |
| *Euvarroa wongsirii* n.sp. | *Apis andreniformis* | Wongsiri *et al.* (1997) |
| *Euvarroa rindereri* De Guzman and Delfinado-Baker | *Apis koschevnikovi* | de Guzman and Delfinado-Ba (1996) |
| *Euvarroa haryanensis* Kapil, Putatunda and Aggarwal | *Apis florea* | Kapil *et al.* (1985) |
| **Laelapidae** | | |
| *Tropilaelaps clareae* Delfinado and Baker | *Apis mellifera* | Atwal and Goyal (1971) |
| | *Apis cerana* | Koeniger and Muzaffar (1988) |
| | *Apis dorsata* | Bhardwaj (1968) |
| | *Apis florea* | Abrol and Putatunda (1995) |
| | *Apis laboriosa* | Delfinado-Baker and Peng (19 |
| *Tropilaelaps koenigerum* Delfinado-Baker and Baker | *Apis dorsata* | Delfinado-Baker (1982) |
| | *Apis cerana* | Abrol and Putatunda (1995) |
| *Tropilaelaps mercedesae* n.sp. | *Apis dorsata* | Anderson and Morgan (2007) |
| *Tropilaelaps thaii* n. sp. | *Apis laboriosa* | Anderson and Morgan (2007) |

## PHORETIC MITES

Astigmata represents the mites having symbiotic association with many species of arthropods. They are the most abundant pests of stored food products, but also inhabit the beehives for the debris. These are the scavengers feeding on pollen, organic debris and fungi and are vectors of bacterial and fungal spores (Eickwort, 1988). A few species of prostigmata have facultative association with honeybees. The species of *Neocypholaelaps* and *Afrocypholaelaps* feed on the pollen of subtropical and tropical trees but are phoretic on adult bees (Haq *et al.*, 2001). *Neocypholaelaps indica* is a common phoretic mite associated with all the major species of honeybees (Nagaraja and Reddy, 1996). It is most abundant on stored pollen and foraging bees. It occurs on more than 34 kinds of flowers and also phoretic on butterflies and other true flies (Ramanan and Ghai, 1984). These mites are known to frequently infest beehives where they may feed on pollen. The tarsonemid mite, *Pseudacarapis indoapis*, feeds on stored pollen and fungal debris in *A. cerana* colonies. However, adult mites are sometimes found phoretically on worker bees. Despite the large number of mites that are occurring as phoretic on adult bees, there is no evidence of significant adverse effect on the general health of the colony.

## PARASITIC MITES

Parasitic mites are economically important as they cause significant loss to bee colonies. These are categorized as endoparasitic as they feed on haemolymph of the tracheae of adult bees and ectoparasitic as they feed on the haemolymph of the developing brood inside the capped cells and also on adult bees. *Acarapis woodi*, *Varroa destructor*, *Varroa jacobsoni* and *Tropilaelaps clareae* are found to be destructive parasites on honeybee colonies. Among these, *A. woodi* is endoparasitic in nature.

## *Acarapis woodi* Rennie

Mites are found to be common in the tracheal system of arthropods. About 15 species of mites are known to parasitize on the tracheal system in the members of Hymenoptera, Orthoptera and Hemiptera (Sammataro, 1995). *Acarapis woodi,* commonly known as tracheal mite, is an endoparasite which inhabits the tracheae and air sacs of adult bees. It was first observed even before this century when *A. mellifera* bees in Europe were dying from unknown disease. This disease was first assumed to be caused by a bacterium, later the causal agent was identified as *Nosema apis,* a microsporidian parasite (Zander, 1909). It was re-investigated and named as *Tarsonemus woodi* and finally as *Acarapis woodi* (Rennie *et al.,* 1921). The damage caused by this mite is commonly called as Acarine or Isle of Wight disease (Hirst, 1921).

The mite has been reported from all over the world. It was confirmed parasitizing on *A. cerana* colonies in India (Singh, 1957) and Nepal (Abrol, 2000), and also on *A. dorsata* from Pakistan (Delfinado-Baker *et al.,* 1989). Atwal and Sharma (1970) reported a loss of 25 to 40 per cent colonies annually due to acarine infestation in India.

## Symptoms

*Acarapis woodi* infests the tracheae that would lead from the first pair of thoracic spiracles of adult bees (Figure 8.1). The infested bees crawl on the ground with disjointed hindwings which are commonly referred to as K-wings. They have distended shining abdomen which generally become sluggish, paralytic and form scattered clusters on the brood. Liu *et al.* (1989) found a secondary coating on the interior part of the trachea of infested bees, which led to insufficient supply of oxygen. The punctures caused by the mites become melanized resulting in dark brown spots on the tracheal wall. These spots become blackened and brittle. In addition, the mites damage the flight muscle fibres. Infested bees sometimes show bacterial infection particularly in their haemolymph and probably this infection may cause more damage than the mites alone.

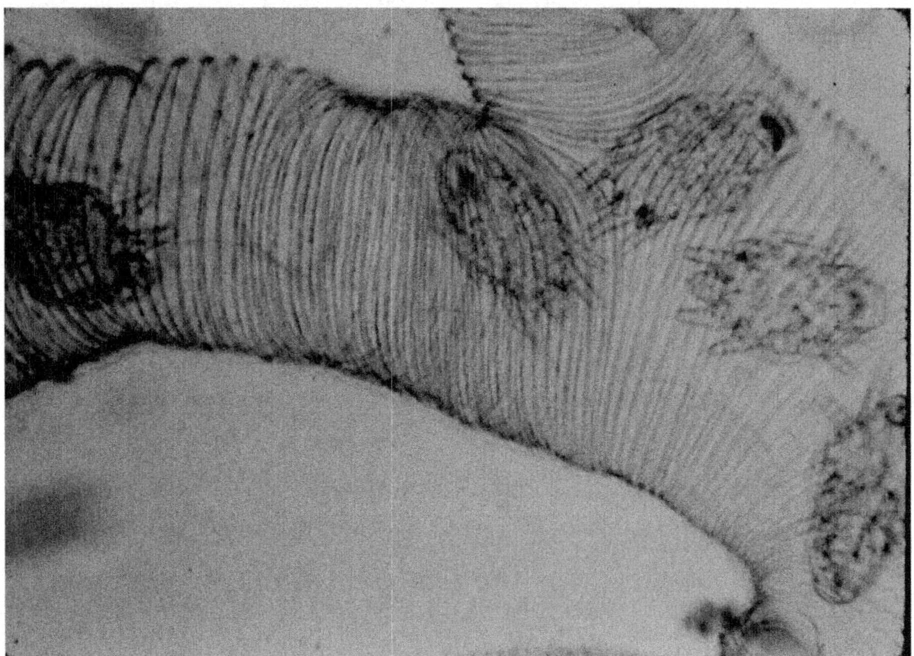

**Figure 8.1**  Infestation of *Acarapis woodi* in the trachea of honeybee

## Diagnosis

*Acarapis woodi* is not visible to the naked eye making the diagnosis of the mite infestation difficult. About 10 suspected bees crawling near the hive with K-winged condition are anaesthetized and pinned on their back to a piece of cork through the thorax. The head and a chitinous ringlike structure around the opening to the thorax are removed with a pair of forceps under microscope. On exposing the trachea

by teasing, various developmental stages of *A. woodi* can be seen. Serological techniques such as enzyme-linked immunosorbent assay have been developed for detecting the acarine mite (Grant *et al.*, 1993). In addition, presence of guanine, a nitrogenous waste product in the excreta of infested bees, is also advocated as a diagnostic tool for mite infestation.

Acarine infestation spreads through emerging mites, robber bees, drifting bees and beekeepers. It is opined that mites are attracted to the openings of the first thoracic spiracle by the puffs of air coming out of bees. Transmission of *A. woodi* from *A. cerana* to *A. mellifera* has also been recorded (Adlakha, 1976).

## Biology

The body of *A. woodi* is oval in shape with a shining smooth cuticle. It has an elongated beaklike gnathosoma with long bladelike stylets. Adult female mites are 120 to 190 $\mu$m long and 77 to 80 $\mu$m wide and males are 125 to 136 $\mu$m long and 60 to 77 $\mu$m wide. The adult gravid female mites enter into the tracheae of worker bees of less than three days old (Gary *et al.*, 1989) and use host cuticular hydrocarbon as cues (Phelan *et al.*, 1991). The life cycle of *A. woodi* has three developmental stages, viz, egg, nymph and the adult (Figure 8.2). Females lay 5 to 7 eggs that hatch in 3 to 4 days. The first egg after laying develops into male and subsequent eggs into females. The nymphs are active feeders and have six legs of which one pair is well-developed and other two pairs are rudimentary. The mite has an apodus nymphal instar present inside the larval skin (Lindquist, 1986). Within the larval cuticle, the mite moves its legs and fixes the opisthosoma dorsoventrally (Pettis and Wilson, 1996). Further it develops into an adult without moulting. The male mites emerge in 11 to 12 days and females in 14 to 15 days.

Mites feed on the haemolymph of bees by piercing the tracheae with sharply pointed styletlike chelicerae by formation of wound.

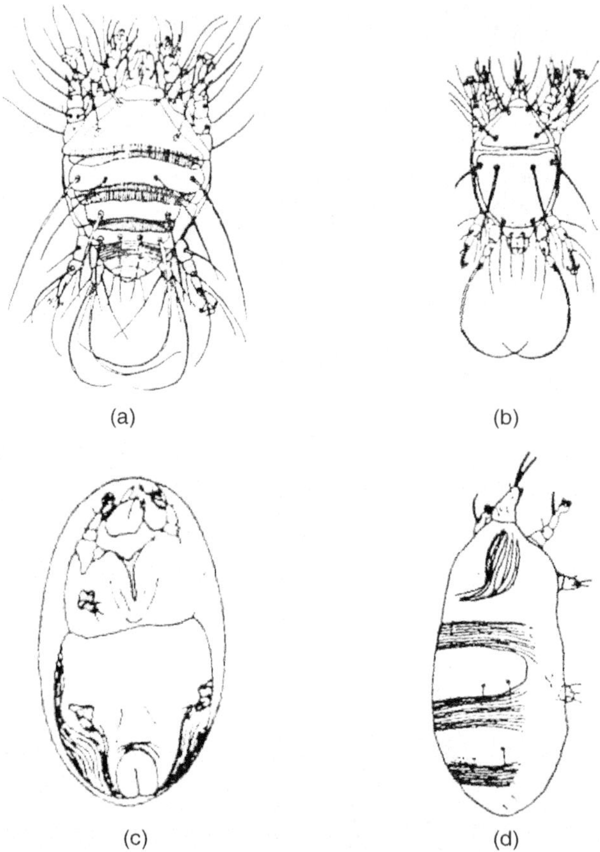

**Figure 8.2** *Acarapis woodi* (a) Adult female, (b) Adult male (c) Larva within its egg shell (d) Nymph (Hirst, 1921)

## Management

***Manipulative methods*** *Acarapis woodi* infests only the adult bees and hence the transfer of brood from infested to healthy colonies is effective in faster growth without loss of brood.

***Breeding for mite tolerance*** The resistance against acarine disease depends on the ability of worker bees to groom the adult mites. de Guzman *et al.* (1998) recorded low level of mite infestation in few bee stocks. Few strains of *A. mellifera* have been reported to have developed tolerance against *A. woodi* (Danka, 2001; de Guzman *et al.*, 1996) and breeding such tolerant stock seems to be a promising method in the elimination of tracheal mites.

***Chemical control*** Management of tracheal mites needs special attention as its development and infestation occurs inside the tracheae of honeybees. Therefore the acaricides used against the mite could reach the tracheae in the form of vapour, but these vapours are lethal to the parasitic mites. Factors such as the concentration of compounds, duration of treatment, colony conditions and temperature contribute to the efficacy of an acaricide. The colonies are generally treated in the beginning of the brood-rearing season. Chlorobenzilate (Folbex) strips are used in the mite-infested colonies during non-honey-flow seasons. These strips are fixed to the top inner cover of the super chamber leaving the strip hanging downwards. These sheets are ignited to smoulder on returning all foragers to the hive by closing the entrance, cracks and crevices. Atwal (1971) recommended 5 to 7 fumigations with chlorobenzilate strips at weekly intervals. Similarly, Shah and Shah (1988) used 1 to 2 strips in the early spring and late summer seasons.

Formic acid could be treated by soaking on absorbent cardboard and placed on the floor of the colony. Sharma *et al.* (1983) recommended 85 per cent formic acid (5 ml/day) continuously for three weeks. Alternatively six weeks of continuous exposure to the vapours of menthol and thymol controls the mite infestation (Kshirsagar *et al.*, 1980, Southwick, 1988). However, in cold conditions, menthol sublimation is not effective as insufficient amount of vapour is released from the crystals. Some vegetable oils are able to disrupt the dispersal of *A. woodi* (Smith *et al.*, 1991). The mixtures of nitrobenzene, methyl salicylate and petrol are also effective but may induce damage to the brood.

## *Acarapis externus* Morgenthaler
## *Acarapis dorsalis* Morgenthaler

*Acarapis externus* is an external mite reported to cause wing loss and malfunction in honeybees. These are found on the neck and posterior tentorial pits on the back of the head and thorax of the bee. Similarly, *Acarapis dorsalis* hides in the scuto-scutellar grooves of the thorax at the base of the wings, propodium and the anterior part of the abdomen. It has many hosts other than honeybees and causes greater infestation during spring and summer. It completes its life cycle in 9 to 10 days (Ibay, 1989). Both the species of mites parasitize only *A. mellifera*.

## *Varroa destructor* Anderson and Trueman
## *Varroa jacobsoni* Oudemans

The genus *Varroa* represents a highly specialized species of obligate ectoparasitic mites that feed on the haemolymph of honeybees. *Varroa destructor* is widely distributed in the colonies of *A. mellifera* and *A. cerana*. *Varroa jacobsoni* parasitizes *A. cerana* in Asia and *A. nigrocincta* in Indonesia (Anderson and Trueman, 2000). Similarly, *Varroa underwoodi* is a parasite of *A. cerana* in Nepal (Delfinado-Baker and Aggarwal, 1987) and *Varroa rindereri* parasitizes *A. koschevnikovi* in Borneo (de Guzman and Delfinado-Baker, 1996).

*Varroa destructor* was considered as *V. jacobsoni* up to its re-classification as separate species (Anderson and Trueman, 2000). It was first detected by the Dutch acarologist Jacobson as a parasite on *A. cerana* in 1904 (Oudemans, 1904).

The studies on phenotypic, reproductive variation and mitochondrial DNA analysis showed that *Varroa* has long been referred to as *V. jacobsoni* which parasitizes different populations of *A. cerana* throughout Asia with a complex of at least two species. Anderson and Trueman (2000) identified haplotypes concealed with the complex of mites infesting *A. cerana*, which have become pests of *A. mellifera* worldwide. The Korean haplotype being a parasite of *A. cerana* in Korea became pests of *A. mellifera* in Europe, New Zealand, the Middle East, Africa, Asia, Canada, and North and South America. The Japan and Thailand haplotype, being a parasite of *A. cerana*, became a pest of *A. mellifera* in Japan, Thailand and North America. The Korean haplotype of *V. destructor* appears to be more parasitic on *A. mellifera* than Japan and Thailand type. The *Varroa* spp. that infests and reproduces in the drone brood of *A. cerana* other than mainland in Asia is *V. jacobsoni* and *Varroa* species which infests and reproduces in *A. mellifera* worldwide is *V. destructor*.

## Symptoms

Bee colonies infested with *Varroa* mites generally become weak and show a spotty brood pattern with punctured capping. Individual brood infested with a few adult mites usually emerge without visible damage. However, they may suffer from malnutrition and loss of the haemolymph. The mites pierce the soft intersegmental tissues of the abdomen and feed on the haemolymph. The bees become stunted with deformed legs and wings that are found crawling on the combs and also on the ground. The individual bees infested with many mites usually become crippled or die. Parasitized pupae would appear to have small pale or reddish brown spots on their normal white bodies. The larvae die in the pre-pupal stage with characteristically raised heads that might be mistaken for sac brood disease (Garg and Kashyap, 1998). The lifespan of infested bees is reduced and the performance of the colony becomes poor.

*Varroa* infestation is associated with the appearance of malformed appendages, shortened abdomen, and reduced weight of bees on emergence (De Jong *et al.*, 1982). The deformity of bees is also due to microbial septicaemia transmitted by *Varroa* (Liu, 1996). *Varroa* is also known to transmit many viruses which cause diseases in honeybees (Ball and Allen, 1988).

## Diagnosis

Examination of sealed brood cells with perforations is an ideal method for diagnosing the *Varroa* mites in the apiary. Mites could be detected by pulling up capped brood cells using a scratcher. The dead mites could also be seen on the bottom board collected on a white paper. Similarly, a plastic sheet smeared with sticky material such as petroleum jelly is placed on the bottom board and the hive is smoked with tobacco for a few minutes. The plastic sheet is removed and the level of mite infestation could be assessed. Special sticky boards have been developed to ease the task of counting mites (Ostiguy *et al.*, 1999). Guanine, the faecal material of *Varroa*, could be seen as white spots on the walls of brood frames in highly infested colonies (Erickson *et al.*, 1994).

*Varroa* mites can be detected by shaking bees in different liquids in a rotatory shaker. The mites along with the debris are placed in a jar filled with 98 per cent alcohol and mites float on the surface on shaking. In ether roll method, hundreds of live bees are collected in a wide mouthed jar by using a short burst of ether (Gruszka, 1988). The bees are rotated in a jar for a few seconds when mites adhere to the surface of the jar. Rapid methods such as washing and

separation of mites from bees through centrifugal force in a closed container at different rotational speed with detergent solution have been developed (Fakhimzadeh, 2000). Pouring such liquid through a coarse mesh would strain out the bees, and the mites can be filtered and counted.

## Biology

Adult female *Varroa* mites are 1.4 to 1.9 mm long and 1.6 to 1.7 mm wide. They are dorsoventrally flattened, pilose, reddish brown, crab-shaped and can be seen with the naked eye (Figure 8.3). Male mites are ovoid, much smaller than females and pale in colour. Several morphological features have made *Varroa* mites as successful ectoparasites. The curved body of the parasite easily fits into abdominal folds of the adult bee. The base of each tarsus is modified into a lobed sucker. The stiff hairs on the ventral side of the body make it difficult for bee to knock off the mites. The female's chelicerae have no fixed digits but the movable digit is a sawlike blade capable of piercing and tearing the integument of the host (Griffiths, 1988). Mites are mainly found on regions of the head and thoracic joints and also on the bee abdomen. The mites remain on these regions, where they can easily penetrate the inter-segmental membrane and gain access to the haemolymph. They prefer to feed mainly on young bees (Hoppe and Ritter, 1989).

The species of *Varroa* are known to mate in capped brood cells in bee colonies. De Jong *et al.* (1982) gave detailed account on the life history of *Varroa*. The life cycle begins with a mated female entering a brood cell few hours before cell capping. The mother mite hides under the larval food at the bottom of the cell for few hours and starts feeding on the brood food during which the bee starts to spin a cocoon. The modified peritremes of mites protrude out of the fluid surface of the larval jelly which enables them to respire (Donze and Guerin, 1997). It releases the mite from larval food to feed on the haemolymph of the host.

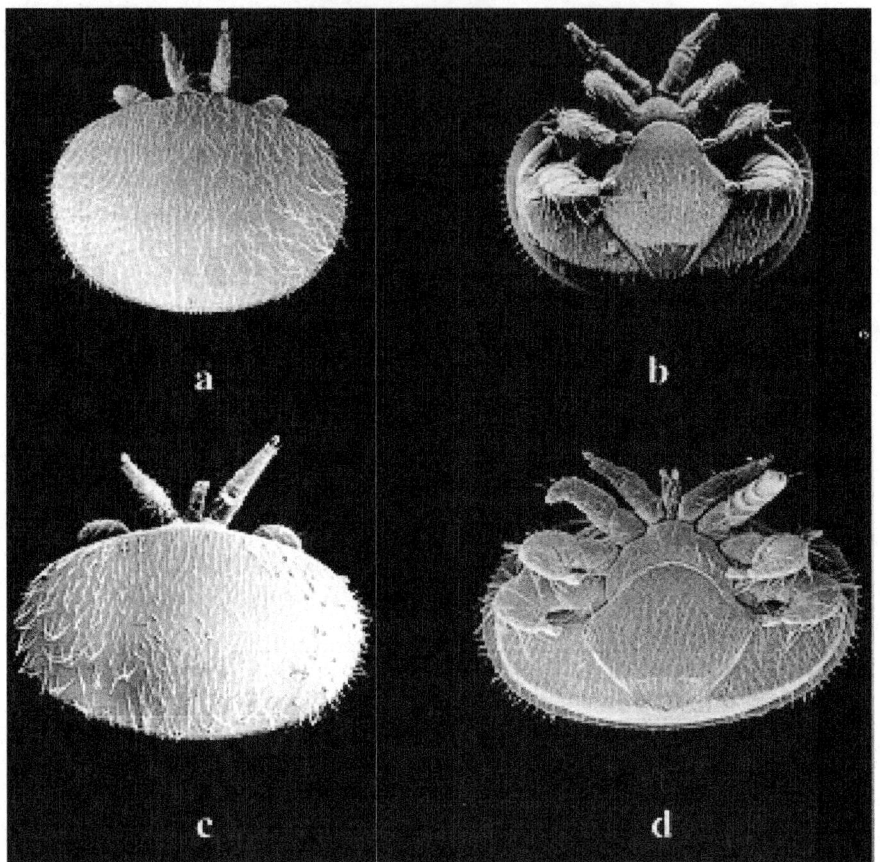


**Figure 8.3**    Dorsal and ventral views of female *Varroa jacobsoni* (a and b) and *Varroa destructor* (c and d)

The mother mite lays the first egg on the cell wall where it is less likely to be damaged upon bees moulting after 60 hours of cell capping (Martin, 1994a). The first egg develops into a haploid male and the subsequent eggs laid develop into diploid females. Mites develop through pharate larva, mobile protonymph, pharate deutonymph, mobile deutonymph, pharate adult and matured adult stages

(De Jong, 1997). The egg hatches into six-legged larva within a day. The larva develops into eight-legged protonymph after two days, which feeds on the haemolymph. It develops into deutonymph followed by adult in the next three days. The nymphs actively feed and develop while pharate instars are in a quiescent stage. The nymphal stages have a number of unusual morphological characters, which are constant throughout the development. The mother mite facilitates her offspring to feed the haemolymph even by pushing away the posterior legs of pre-pupal stage of bees (Donze and Guerin, 1994). The developed female mites mate with males in the capped cells on maturation. The total developmental period is 8 to 10 days for females and 6 to 7 days for males.

The nature of the genital opening, movable digit of the chelicerae and the forms of peritremes and tritosternum are the distinguished characteristics of the adult males (Delfinado-Baker, 1984). The male mates with emerging mites in succession by injecting sperms into a pair of induction pores in females located between the base of the III and IV pair of legs by using specially modified chelicerae. Unmated females produce only male offspring. The chelicerae of the males are modified for sperm transfer and therefore males cannot feed, and die soon. The mother and mated young female mites come out through the emerging adult bees. The mites emerge from the brood cells without killing the bees, but several mites in a cell would kill the host. The mother mite may invade a second cell and is known to complete up to eight cycles of reproduction under favourable conditions (Martin and Kemp, 1997). An infested worker and drone bees may act as the host for 5 and 12 mites, respectively. The lifespan of the adult mites is about 2 months in summer and 5 to 8 months in winter.

*Varroa jacobsoni* reproduces only in the drone brood of *A. cerana*. In the absence of drone brood, the mother mites may invade worker cells but they do not lay eggs due to non-activation of oogenesis. They feed on the developing pupae of workers. In some parts of Asia, the genotypes of *V. jacobsoni* may invade *A. mellifera* colonies at low levels but they do not reproduce either on worker or drone brood (Anderson, 1994).

## Management

***Manipulative methods***   One of the earliest methods of management of *Varroa* is trapping mites from the drone brood. Mites are found more often on drone brood than on worker brood with average differences between 5 and 12 folds (Calderone and Kuenen, 2001). The hand removal of sealed drone brood from freshly infested colonies suppress the development of mite population. Creating the colonies broodless by caging the queen for three weeks is also followed but with limited success. Dusting of wheat flour on bee combs at an interval of 10 days reduces the level of *Varroa* infestation (Loglio and Penessi, 1991). The fine dust of wheat interferes with movement of the mite by adhering to the ambulacrum. The separate application of glucose powder and finely ground white sugar powder (5–15 $\mu$m) on bees leads mites to lose balance and fall on the bottom board and later they die due to starvation (Fakhimzadeh, 2001).

***Biological control***   Biological control attracts attention either as a partial or total alternative to chemicals in an integrated pest management system. Several spores of entomopathogenic fungi have been found to infect *Varroa* mites under laboratory conditions. Meikle *et al.* (2007) recorded a significant increase in the percentage of *Varroa* mites infected by *Beauveria bassiana (Bals.criv.)* Vuill under field conditions. Mites are allowed to move on the culture

isolates of the fungus, *Hirsutella thompsonii* Fisher, which infects the mite's suckers. Similarly, application of *Metarhizium anisopliae* Metchnikoff spores kills mites (Roberts and Leger, 2004). The fungal spores do not damage brood and adult bee population. Dusting the colonies with a formulated product of *M. anisopliae* (Bio-Blast®) (about $4\times10^9$ visible spores/g) at 10 days interval increased mite mortality (Kanga *et al.*, 2003). In addition, various predatory mites, parasitoids and entomopathogens have been recommended as potential bioagents for the control of *Varroa* spp. (Chandler *et al.*, 2001).

***Breeding for Varroa tolerance*** Development of resistance to wide range of pesticides is a well-known phenomenon among the mites. The resistance of mites to pyrethroids was detected in Italy around 1991 followed by Switzerland, Slovenia and southern France (Martin, 2004). Breeding tolerant colonies is one of the remedies to overcome the *Varroa* problem. *A. cerana* perform auto-grooming and group cleaning behaviour in removal and killing of *Varroa* mites (Peng *et al.*, 1987). Similarly, Koeniger and Muzaffar (1988) reported that *A. dorsata* workers are able to injure the mites, which leads to the reduction in the mite's lifespan. The process of opening mite-infested brood cells and removal of the parasitized brood disturbs the reproduction of mites. The ability of bees to manage the mites seems to be a component of regulation between the host and its parasite.

Hygienic behaviour reduces the level of mite population in colonies that require no or less chemical treatment (Boecking and Spivak, 1999). The hygienic colonies uncap and remove the developing mite offspring by interrupting the reproductive cycle. Boecking and Drescher (1992) found greater rate of removal of mites (55 per cent) on the introduction of two mites per cell compared to one mite (29 per cent). The capping of infested brood cells are opened by the bees, but later reseal with wax without removing the brood. Most of the adult females that escape from the brood cells could invade other brood cells (Boecking, 1992). Shortening of pupal period by at least six hours results in greater mortality of mites (Bienefeld, 1996). Breeding for high degree of heterozygosity may be helpful in producing colonies with short-capped period (Büchler and Drescher, 1990).

Harbo and Hoopingarner (1997) bred bees that maintained low mite levels, as they appeared to have low reproductive success on worker brood. They found the trait to be heritable in bees and called it as suppression of mite reproduction trait (SMRT). Mites enter the worker brood cells without reproduction and occasionally produce only males or progeny too late to mature (Harbo and Harris, 1999). Queens selected for this trait had low mite fecundity (Table 8.2). However, the mite population was not suppressed in the first cycle of brood when a queen with genes for SMRT was introduced but took at least six weeks to suppress mite reproduction (Camazine, 1986). Calderone and Kuenen (2001) reported that the cell type and larval sex affect

**Table 8.2**   Heritability of colony characteristics against *Varroa destructor* based on sibling analysis (Harbo and Harris, 1999).

| Characteristics | Range | $h^2 \pm$ SEI |
| --- | --- | --- |
| Suppression of *Varroa* reproduction | 10.66–49.33 | $0.30 \pm 0.55$ |
| Hygienic behaviour | 4–91% | $0.65 \pm 0.61$ |
| Physical damage to mites | 4–35% | $0.00 \pm 0.45$ |
| Capped period (*h*) | 268–290 | $0.89 \pm 0.59$ |
| Proportion of *Varroa* in brood | 39–82% | $1.24 \pm 0.49$ |
| Mites per 100 cells of brood | 4–30.5 | $0.28 \pm 0.56$ |
| *Varroa* population | 534–3389 | $0.17 \pm 0.52$ |
| Mites per 1000 bees | 26–198 | $0.01 \pm 0.46$ |

the level of mite infestation. Similarly, the mite prevalence values are greatest in drone larvae reared in drone cells followed by drone larvae reared in worker cells, worker larvae reared in worker cells and worker larvae reared in drone cells.

Breeding colonies which are resistant to mites is possible by selection of bee stock that detects and removes mite-infested brood effectively. Brückner (1975) reported that the capacity of bees in colony thermoregulation could be affected by inbreeding. A good adaptation of the colony on the availability of food resources becomes increasingly important in *Varroa* tolerance (Büchler, 1998).

***Chemical control*** The effective chemical treatment would not only kill the phoretic mites on adult bees but also their offspring in capped cells. More than 120 chemicals are known to reduce *Varroa* menace in honeybee colonies, but none of them have claimed cent per cent effectiveness (Garg and Kashyap, 1998). Phenothiazine was the commonly used acaricide in eastern Europe, which proved rather ineffective. One of the most successful fumigants is bromopropylate (Isopropyl-4, 4´-dibromobenzilate) sold under different trade names that include Folbex VA, Folbex VA Neu, Folbex Forte, Neoron and Varroatex. The fumigants are burnt directly in the hive or in a smoker with an extra long nozzle. Ritter *et al.* (1984) recommended treating eight colonies together using eight strips of Folbex by giving each colony eight puffs of smoke at an interval of five seconds. Fluvalinate sold as Apistan® and Amitraz® are contact acaricides that show considerable *Varroa* control. Plastic strips impregnated with the active ingredient of these acaricides are placed in the brood nest and bees spread them through their routine activities. Application of two strips per hive for 6 to 8 weeks is effective in heavily mite-infested colonies. Both are non-toxic with limited or no traces of residues in honey. However,

these acaricides do not reach the capped brood cells where development of mites takes place.

The systemic acaricide reported to be successful is Coumaphos (Perizin®). The weight of the active ingredient of Perizin required is 32 mg/colony. The liquid formulation of Perizin is sprinkled on the bee combs. Though single treatment is quite adequate, two applications are highly effective. Similarly, application of Coumaphos (Check mite® or Bayer strips) kills the mites. These chemicals are impregnated in plastic and strips are normally hung between the frames in the hive.

About 150 essential oils or their components have been tested against *Varroa* with varied effectiveness. Many of the tested oils have been proved effective in laboratory experiments and field trials. Ruffinengo *et al.* (2005) reported that *Acantholippia seriphioides* Griseb mould has greater acaricidal effect causing high mortality of mites at low oil concentration. The main components of this oil are thymol, carvacrol and orthocimene. Thymol, a volatile monoterpenoid constituent of thyme, *Thymus vulgaris* Linn. is widely used to control *Varroa*. Two thymol-based fumigants, Api LifeVar® (Imdorf *et al.*, 1999) and Apiguard® are available in Europe against *Varroa* mites (Mattila *et al.*, 2000). ApiLifeVar also incorporates minor constituents of menthol and camphor. Treatment with either of the products provides an effective control. Application of three tablets of ApiLifeVar and two strips of Apiguard was also recommended. Similarly, pure thyme (15 g) dissolved in ethanol (20 ml) is poured into a viscous sponge and is kept on brood combs for evaporation of alcohol. Mite-A-Thol®, an extract from the plant, *Mentha arvensis* Linn., is partially effective against *Varroa* though temperature-dependent. Kshirsagar *et al.* (1980) recommended the fumigation of menthol strips

(300 mg menthol) and recorded no toxicity to honeybees and contamination of hive products.

An alternative *Varroa* control is the application of organic acids. Lactic acid has an efficiency of killing more than 80 per cent mites in broodless colonies. The colonies could be sprayed with 5 to 8 ml of lactic acid (15 per cent) per comb and this is a highly convenient method to treat small apiaries. Oxalic acid (30 g/l of water) in combination with lactic acid and oxalic acid alone could also be applied (Charriere and Imdorf, 2002). Formic acid fumigation is advantageous in indoors as there is a high degree of control over ambient environmental conditions to which colonies are exposed. During field application, volatilization of formic acid is affected by a number of environmental variables which in turn fluctuate the concentrations widely. Biannual treatment of formic acid was found to be effective for the control of *Varroa* (Satta *et al.*, 2005). MiteAwayII® are the pads developed as a delivery system for formic acid. Treatment with formic acid does not affect the colony activities (Sharma *et al.*, 1994). In contrast, it could cause reduction in brood-rearing activity and affect the physiology of the brood and young worker bees (Bolli *et al.*, 1993).

## *Varroa rindereri* De Guzman and Delfinado-Baker
## *Varroa underwoodi* Delfinado-Baker and Aggarwal

*Varroa rindereri* is a parasite of red honeybee, *A. koschevnikovi*. It was recorded from *A. cerana* colonies in Nepal during 1987 (Delfinado-Baker and Aggarwal, 1987). It resembles to *V. jacobsoni* but is smaller in size. The main spiracles that possess a conspicuous tube allows the mite to respire from the food of the host (Oldroyd and Wongsiri, 2006).

*Varroa underwoodi* is an ellipsoidal, light brown ectoparasite whose females are 0.780 mm long and 1.168 mm wide.

The males are round and weakly sclerotized with long hairs on the sides of the body. It infests domesticated honeybees, *A. nigrocincta* and *A. nuluensis* (Otis and Kralj, 2001). The mites are known to feed on the bee brood.

## *Euvarroa sinhai* Delfinado and Baker
## *Euvarroa wongsirii* n. sp.

Delfinado and Baker (1974) described *Euvarroa sinhai* as an ectoparasite of *A. florea*. The female mite is brown, broadly pear-shaped and setaceous without fixed cheliceral digit. The protonymphs, deutonymphs and adults feed on brood and adult bees (Mossadegh and Birjandi, 1986). The adult female mites leave the brood cells and become phoretic on adult bees. However, it is not known as to when they are able to feed on haemolymph from the adult bees (Akratanakul and Burgett, 1976). Its life cycle is almost similar to that of *Varroa* and also infests the colonies of *A. mellifera*.

*Euvarroa wongsirii* infests the drone brood of *A. andreniformis* and is broader and slightly triangular in shape. *Euvarroa haryanensis* Kapil, Putatunda and Aggarwal was found associated with *A. florea* in India and subsequently on *A. mellifera* (Kapil *et al.*, 1985).

## *Tropilaelaps clareae* Delfinado and Baker

The species of the genus *Tropilaelaps* are commonly seen as members of the mammalian parasites. It is an old-world tropical genus that has been evolved away from its presumed predatory ancestry to become a parasite on the brood of honeybees (Woyke, 1984b). Among the species of *Tropilaelaps, T. clareae* has been originally recorded on field rat from the Philippines (Delfinado and Baker, 1961). It has been a native parasite of the giant honeybee, *A. dorsata*

(Bhardwaj, 1968) and distributed in Asia, from Iran in the northwest to Papua New Guinea in the southeast (Ellis and Munn, 2005). Delfinado Baker (1982) recorded *T. clareae* on *A. cerana* colonies from Pakistan and Burma.

*Tropilaelaps clareae* (Figure 8.4) is an ectoparasitic mite that exploits the brood of honeybees possibly by surviving during broodless periods (Nagaraja, 1998b). It is considered to be the most important limiting factor on the development and expansion of beekeeping with *A. mellifera* in Asia (Nagaraja, 2000). The mites are found around the head, thorax and abdomen of the adult bees. The shortened life cycle followed by a brief stay on adult bees encourages faster growth of mite population.

## Symptoms

*Tropilaelaps clareae* possesses predatory characters and its parasitic mode could be considered as a secondary adaptation (Aggarwal, 1990). But these observations are contradicted by Rath *et al.* (1991) who reported that the mite is morphologically and ecologically adapted for phoretic behaviour that distinctly qualifies for ectoparasitic mode of life. The bee colonies infested with *T. clareae* show a scattered brood pattern. The brood cells have sunken capping and the adult mites can be seen often running on the combs. The infested pupae with deformed and mutilated wings and stunted abdomen of the worker bees could be seen near the hive.

*Tropilaelaps clareae* infests both worker and drone brood of *A. mellifera*. Its mean infestation varies from 3.00 to 7.00 per cent in worker brood and 4.00 to 6.32 per cent in drone brood of *A. mellifera* (Nagaraja, 1998b). Woyke (1994) recorded a mite infestation of 54 per cent in the brood and 1.5 per cent in adult bees. The number of mites are found to be higher in worker than drone brood cells.

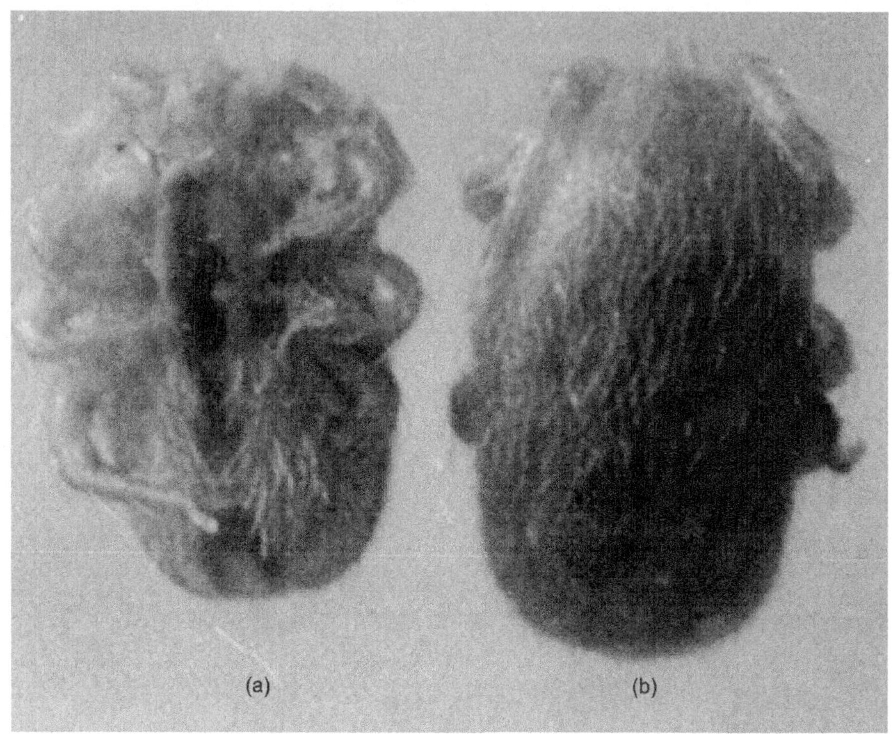


**Figure 8.4**    (a) Ventral and (b) dorsal views of adult *Tropilaelaps clareae*

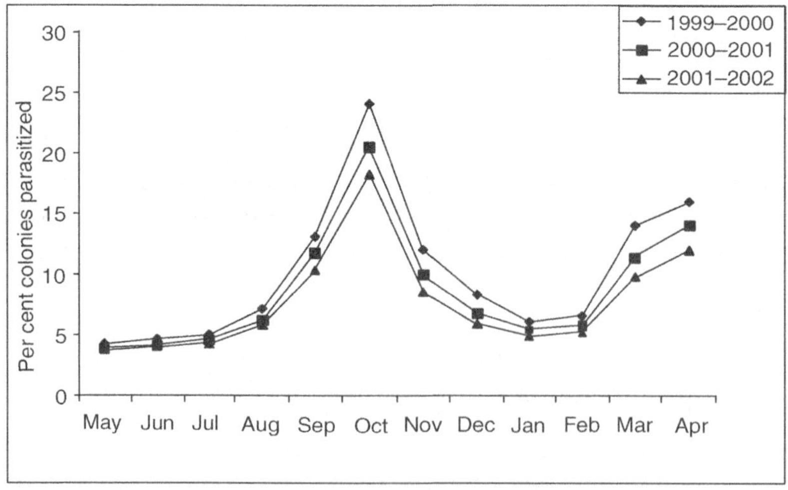


**Figure 8.5**    Parasitization of *Tropilaelaps clareae* in the colonies of *Apis mellifera*

About 60 per cent of worker cells and 45 per cent of drone cells hadonemite each followed by two mites in 25 to 35 per cent of cells. Only a few brood cells had 5 mites each. Similarly, the number of adult mites and nymphs in an infested brood cell ranged from 1 to 15 (Garg and Kashyap, 1998). Atwal and Goyal (1971) recorded about 50 per cent brood mortality in late larval and pupal stages due to mite infestation. A peak mite infestation was recorded during two seasons annually, one in September–November and again in March–April. However, the infestation was recorded low during dearth period (Figure 8.5).

## Diagnosis

Most of the techniques evolved to diagnose *Varroa* mite could also be followed for *T. clareae*. The most reliable method is to open a large number of sealed brood cells and examine the nymphs and adult mites. Mites that die naturally remain dislodged and are collected by placing a thick sheet of paper on the bottom board. The debris collected from infested colonies is placed in a jar filled with alcohol and mites may float on the surface of the jar on shaking.

## Biology

*Tropilaelaps clareae* reproduce in the sealed cells of worker and drone brood of *A. mellifera* and *A. dorsata.* The adult females are 1.0 mm long and 0.550 mm wide, elongated and light reddish brown in colour (Figure 8.4). The mouthparts of female are stubby and suitable for piercing the soft tissues of larvae and pupae and are sufficiently protected from the cleansing activity of honeybees. Males are less sclerotized and their chelicerae are toothed. Several eggs mature almost at the same time. Therefore, the abdomen of the gravid female is bulged at the time of entering the cell. The gravid female mite enters into the cells of the bee brood shortly before sealing and crawl under the larvae by hiding in the larval jelly. After capping, the bee larvae eat the remaining larval jelly and then the mites may climb up and feed the bee larva regularly. After capping, the female mite lays all the eggs within two days (Woyke 1989).

The first egg after laying hatches into a male larva and feeds on the soft abdominal segments of the bee larva. Later, the mite larva metamorphoses into the protonymph followed by the deutonymphal stages. The deutonymphs are oval and resemble the adults except in chitinization. They feed on the bee pupae regularly and develop into adult males in 7 days (Nagaraja and Rajagopal, 2001). Similarly, the succeeding eggs laid by the mother mite hatch into female larvae in one day and reach the protonymphal and deutonymphal stages after 2 and 7 days, respectively. All the offspring mature before the worker bee emerges from the cell. They feed on the abdominal segments of bee larvae and pupae and the female mite may mate with the matured male in the capped cells and come out along with emerging adult bees in 8 days. However, the developmental duration varies with geographical regions (Table 8.3).

**Table 8.3**   Developmental period of female *Tropilaelaps clareae* on *Apis mellifera*

| Developmental period (days) | | | | Total period (days) | Reference |
|---|---|---|---|---|---|
| Egg | Larva | Proto-nymph | Deuto-nymph | | |
| 0.30–0.40 | 0.30–0.50 | 1.70–2.00 | 3.00–3.80 | 5.3–6.7 | Woyke (1987c) |
| 0.96 | 1.00 | 2.16 | 3.86 | 7.98 | Nagaraja and Rajagopal (2001) |
| 1.50 | 1.85 | 2.11 | 3.75 | 9.21 | Kitprasert (1984) |

Male mites usually do not die in the brood cells and emerge together with bees. Woyke (1996) recorded the survival of male mites on pupae and adult bees for two weeks. However, they survive for a month when they are provided with the brood. Adult mites survive for a few days on the bottom board of the hive (Rath *et al.*, 1991). Contrary to *Varroa*, *T. clareae* enters the nearby cells within two days after emerging (Woyke, 1987a). It is possible that, during one reproductive cycle of *Varroa*, *T. clareae* completes almost two cycles. Hence, its population in bee colonies increases in geometrical proportion than *Varroa* population (Woyke, 1987b).

Woyke (1994) studied the mating behaviour of *T. clareae* in detail using stereomicroscope. Mating takes place in a cell with a young bee ready to emerge and occasionally outside the sealed brood cells. The mature male and female mites initially touch the first pair of legs of each other. The male jumps on to the dorsum of the female by stretching its legs forward. The male elicits a kind of sexual stimulus on females by tapping the tarsi. The male vibrates possibly as a prelude to secrete a spermatophore. The mating is carried out by podospermy where the sperm material is transferred to the spermatheca through the two external openings located between coxal segments of III and IV pairs of legs by means of spermatodactyle (Rath *et al.*, 1991). This indicates that the male mites mate with many females similar to that of *Varroa* (Ifantidis and Rosenkranz, 1988). However, Woyke (1994) recorded multiple mating in both newly emerged and older female mites.

The successive reproduction of *T. clareae* in the sealed brood cells is attributed to non-disturbance of worker bees. Similarly, the longer post-capping period of the host compared to the shorter developmental period of the parasite made its successful reproduction in the brood of *A. mellifera*. *T. clareae* multiplies in the brood of *A. dorsata* as they are found to migrate after brood season by leaving the mite-infested brood. In addition, *A. dorsata* and *A. laboriosa* develop a unique mechanism to prevent the attack of mites. They do not open the mite-infested pupae in the capped cells. Therefore, the parasite which kills the pupae do not reach the adult bees or other brood cells (Woyke, *et al.*, 2004).

## Management

***Manipulative methods*** *Tropilaelaps clareae* does not survive for long periods in bee colonies without brood. Creating broodless conditions for 2 to 3 weeks by caging the queen reduces the mite population (Woyke, 1985). Dusting of sugar powder (12 g and 15 g /colony) on mite-infested adult bees of *A. mellifera* had recorded a fall in mite population by 85.15 per cent and 91.20 per cent respectively (Figure 8.6). Application of fine sugar powder dislodges the mites from the body surface of bees and they die on the bottom board due to starvation (Nagaraja and Rajagopal, 2003a).

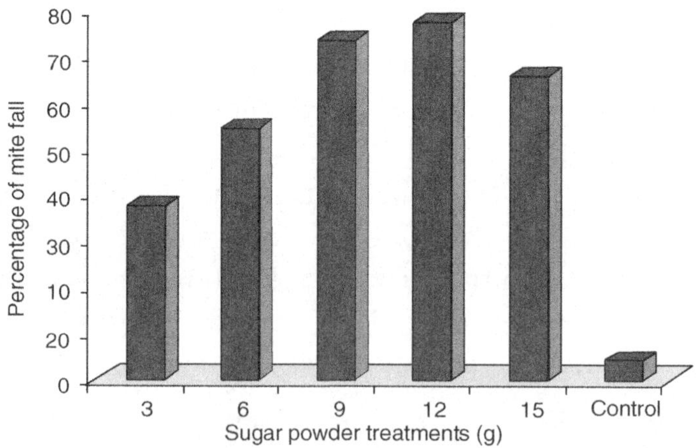


**Figure 8.6**    Efficacy of sugar powder dusting on the fall of *Tropilaelaps clareae*

This treatment is not only effective during brood-rearing season but also during honey-flow season. In addition, sugar dusting stimulates brood rearing as observed in feeding sugar-

syrup-treated colonies during the dearth period. This method is also useful in dislodging of phoretic mites on honeybees.

***Breeding for mite tolerance***  Koeniger *et al.* (2002) found a number of capped brood cells in combs deserted by *A. dorsata* colonies as a mechanism for mite elimination. Examination of brood cells of *A. dorsata* infested with *T. clareae* reveal that the bee pupae are not killed even if the cell is invaded by 2–3 mites. However, they are killed on invasion by many mites (14 mites/cell). Hygienic colonies of *A.dorsata* remove a part of introduced pin-killed brood within a day. The rate of removal was high on cell capping pinned with a harder pin. The harder pin causes greater damage to pupae which leads to oozing out of larger amount of haemolymph. This is the main cause for the faster removal of the brood (Woyke *et al.*, 2004). *A. dorsata* workers groom and kill the mites outside sealed brood cells. These are also found to remove the introduced mites through auto-grooming and allo-grooming behaviour (Büchler *et al.*, 1992).

The workers of *A. mellifera* remove the mites through exhibiting three types of behaviour such as self-cleaning, nest-mate cleaning and group cleaning (Figure 8.7). In the self-cleaning behaviour, bees perform vigorous body vibration. Mites hiding in the thoracic and abdominal intersegmental regions are thereby dropped on to the bottom board. The groomer also responds to the nest mates by raising up its thorax and abdomen. On locating the mite, the nest mate picks up the mite and drops on to the bottom board (Nagaraja *et al.*, 2003). Cleaning by nest-mate behaviour draws the attention of a few other bees in the vicinity, which behave in the same pattern as the nest mate, and this is normally referred to as a group-cleaning behaviour. About 93 per cent of the introduced mites are removed from the colonies exhibiting self (27 per cent), nest-mate

(52 per cent) and group-cleaning (14 per cent) behaviour. The bees detect and remove the mites through olfactory–visual mechanism.

Hygienic behaviour is one of the selective traits for queen breeders. The hygienic behaviour tested in the colonies of *A. mellifera* by introducing pin-killed brood reveal that the hygienic colonies remove mite-infested brood at a faster rate. Among the daughter colonies produced from a hygienic stock, a few daughter colonies are hygienic against *T. clareae*. About 52 per cent and 43 per cent of the introduced mites are removed within 5 minutes and 90 minutes respectively. However, 5 per cent of the introduced mites are not successfully removed from the brood cells. It also takes more time for the worker bees to detect and remove the small papers placed in the brood cells (Figure 8.8). Therefore, selective breeding programmes could be undertaken to produce mite-tolerant bee colonies.

***Chemical control*** Most of the acaricides used on *Varroa* may also be effective against *T. clareae*. Before application of any chemicals in bee colonies, the short stay of the female mites outside the brood cell would be considered into account. Chlorobenzilate (Folbex) strips treated at 6 and 12 weekly intervals are not effective (Atwal and Goyal, 1971; Laigo and Morse, 1969). Failure of these treatments may be due to successive development of the mites before the application of acaricides (Woyke, 1987c). Thus, any acaricide applied against the mite must be active in the hive at least for 2 to 3 weeks.

Nyein and Zmarlicki (1982) recommended removal of brood by caging the queen for at least three weeks followed by subsequent fumigation with phenothiazine. Woyke (1987d) found the fluvalinate (Amitraz®) as an effective acaricide. Similarly, Burgett and Kitprasert (1990) evaluated the fluvalinate (Apistan®) under tropical conditions and

reported it to be a very effective fumigant against *T. clareae*. Garg and Sharma (1988) recommended application of an effective dose (200 mg/frame) of sulphur for four weeks. Continuous fumigation of 85 per cent formic acid for three weeks was found to be effective in killing the mites (Garg *et al.*, 1984).

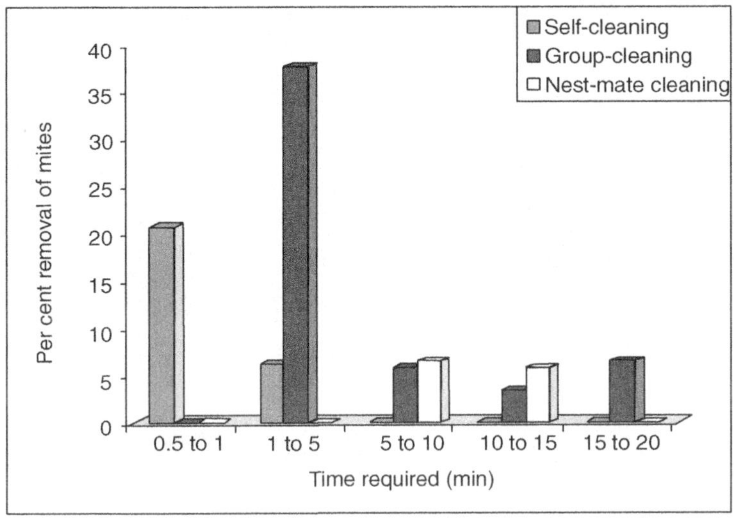


**Figure 8.7**     Efficiency of *Apis mellifera* in removing alien *Tropilaelaps clareae*

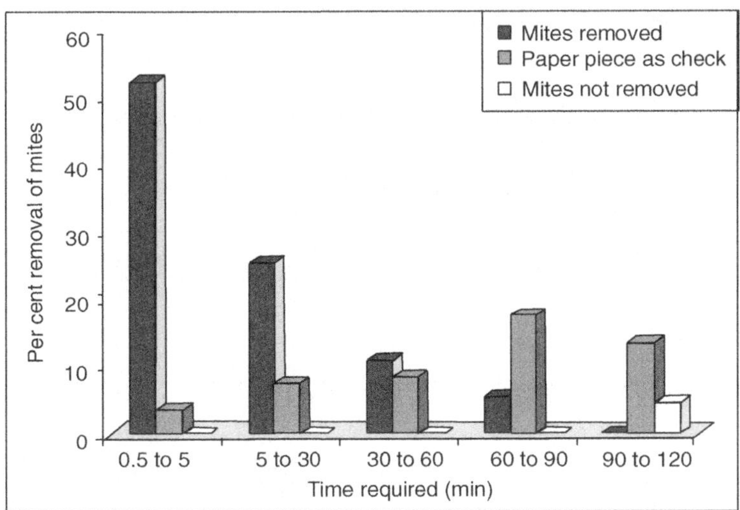


**Figure 8.8**     Efficiency of *Apis mellifera* in removing *Tropilaelaps clareae* from worker brood

Natural plant products offer desirable alternatives to synthetic chemicals as they have low mammalian toxicity and less environmental contamination. Azadirachtin is an important terpenoid present in the neem seeds and has been identified as an active component with repellency and antifeedant activity against many insects. The different concentrations of neemarin 1500 (azadirachtin 0.15 per cent), a commercial neem product, has a significant effect on mortality of *T. clareae* in *A. mellifera* colonies. Greater mite mortality (90 per cent) was recorded on treatment with 200 ppm and minimum mortality (45 per cent) on application of 50 ppm (Figure 8.9). The greater mite mortality was attributed to high repellency and feeding deterrence of azadirachtin. (Nagaraja and Rajagopal, 2003a). Garg and Sharma (1988) reported that thymol (10 g/colony) was found to be effective against *T. clareae* mite population.

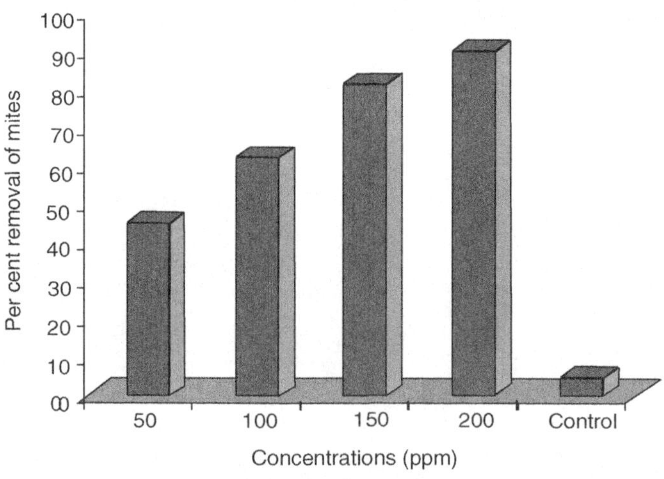


**Figure 8.9** Efficacy of neemarin on the mortality of *Tropilaelaps clareae*

## *Tropilaelaps koenigerum* Delfinado-Baker and Baker

*Tropilaelaps koenigerum* has been reported as a new species of ectoparasitic on *A. dorsata* in Sri Lanka (Delfinado-Baker

and Baker, 1982). It has been found to occur on *A. cerana* and *A. mellifera* (Abrol and Putatunda, 1995) and *A. laboriosa* (Delfinado-Baker *et al.*, 1985). Mites are oval and light brown coloured, and their dorsal plate is highly sclerotized.

## Tropilaelaps mercedesae n.sp.
## Tropilaelaps thaii n.sp.

*Tropilaelaps mercedesae,* which encompasses haplotypes of *T. koenigerum,* parasitizes *A. dorsata* in mainland of Asia and Indonesia except Sulawesi Island. It also parasitizes *A. mellifera* in these regions. *Tropilaelaps thaii* parasitizes *A. laboriosa* in the Himalayan regions (Anderson and Morgan, 2007).

## OTHER SPECIES OF MITES

*Melittiphis alvearius* Berlese (Acari: Laelapidae) has been found to feed on queen bee of *A. mellifera* in the United Kingdom, which was introduced from New Zealand (Cook and Bowman, 1986). This species has been noticed on the debris collected from the hives of *A. mellifera* in Czechoslovakia (Samsinak *et al.*, 1978). *Pyemotes herfsi* Oudemans (Acari: Pyemotidae) is an ectoparasitic mite found on *A. cerana* in India (Dinabandhoo and Dogra, 1980). It interferes with normal functioning of mouthparts and other activities of bees in the colony.

9

# Insect Pests and Predators

Honeybee colonies are attacked by varieties of insect pests which cause serious losses to beekeeping in tropical and sub-tropical regions of the world. The wax moths, hive beetles, ants and wasps are the major insect pests and predators. Honeybee combs are vulnerable to two species of wax moths, the greater wax moth, *Galleria mellonella*, and lesser wax moth, *Achroia grisella*. Both the species may occur naturally or might be introduced by man to beekeeping regions of the world (Haung, 1984). The insect pests infesting honeybee combs are presented in Table 9.1.

## WAX MOTHS

### *Galleria mellonella* Linn.

The occurrence of *G. mellonella* on bee colonies has been reported from the early days of Aristotle (384–322 BC). It is one of the most important enemies of bee colonies causing serious damage particularly in weak colonies. In India it is observed infesting throughout the year in both higher and lower altitudes, but peak infestation is recorded during May–September. Combs of all the species of *Apis* are freely attacked by wax moths.

**Table 9.1**   Common insect pests infesting honeybee combs

| Order/Family | Common name | Scientific name | Status |
| --- | --- | --- | --- |
| **Lepidoptera** | | | |
| Pyralidae | Greater wax moth | *Galleria mellonella* Linn. | Major |
| | Lesser wax moth | *Achroia grisella* Fab. | Minor |
| Sphingidae | Hawk moth | *Acherontia styx* Westwood | Minor |
| **Coleoptera** | | | |
| Nitidulidae | Small hive beetle | *Aethina tumida* Murray | Major |
| | Large hive beetle | *Oplostomus fuligineus* Olivier | Minor |

## Nature of Damage

The extent of wax moth damage may vary in honeybee colonies through geographical regions. The dearth season results in food scarcity to bees. Under such conditions, the bee colonies become weak and are attacked by wax moth. The combs are infested in store, as well as in hives unoccupied by bees. The wax moth larvae burrow into the comb and reach the midrib region by producing silken tunnels along with their excreta. They live in the silken tunnels and feed on the propolis, pollen and beeswax in the combs. During severe infestation, the combs are being found to be destroyed (Figure 9.1) which causes the bee colonies to abscond. Mahindre (1983) reported about 90 per cent infestation of wax moth in the combs of *A. dorsata*. However, Brar *et al.* (1985)

recorded 16 to 19 per cent infestation in *A. mellifera* colonies in north India.

**Figure 9.1**  Honeybee comb damaged by the larvae of greater wax moth, *Galleria mellonella*

Occasionally, the bee pupae are exposed outside after removal of cell capping by wax moth larvae, and this condition is known as bald brood. One of the causes for this deformity in developing bee pupae is a result of deposition of wax moth excreta in the pupal cells. Similarly, worker bees also chew and remove a part of the capping indicating the direction of wax moth infestation. Studies showed that the infestation of wax moth varied with different seasons in the colonies of *A. mellifera* (Figure 9.2). Greater infestation was recorded from June to August and lower infestation during high brood-rearing seasons (Nagaraja and Rajagopal, 2003b). Similarly, Viraktamath (1989) recorded peak wax moth infestation during May–August, which coincides with the dearth period of honeybees under south Indian conditions.

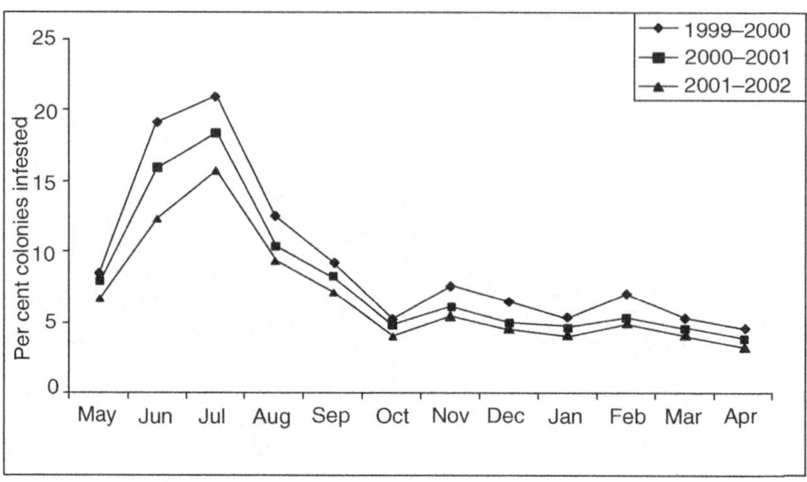


**Figure 9.2**    Infestation of greater wax moth, *Galleria mellonella,* in the colonies of *Apis mellifera*

## Biology

The life stages of *G. mellonella* consist of egg, larva, pupa and adult (Figure 9.3). The adult moths are brownish grey in colour and 12–16 mm long. The males are slightly smaller than the females and can be distinguished by the shape of the outer margin of the forewing, which is smooth in females but notched in males. They mate after emergence and about 200 to 300 eggs are laid in batches in two days. The eggs are laid most frequently in the cracks and crevices between supers, bottom board and on the frames when combs are fully occupied by bees. The eggs are creamy white, almost elliptical (varying in shape depending on the number of eggs laid in a spot) and hatch in 8 to 10 days. The young larvae are active and often seen on the combs and also moving in the cracks and crevices of the hive. The fully grown larvae are cylindrical, smooth and greyish in colour. They try to make burrows into the wax comb immediately after emergence and create small tunnels between the cells and midrib of the comb through silken strands of web.

The tunnels are enlarged as larvae grow, and matured larva are about 28 mm long (Burges, 1978). The growth of the larvae depends upon several factors, including quantity and quality of food and prevailing temperature. The duration of larval period may vary from one to five months.

The larvae prefer to live in darker combs rather than the honey-extracted combs and confine mostly to the midrib and base of the cells in empty combs. The fully grown larva spins a dense silken cocoon, which usually attaches firmly to the hive parts. The duration of pupal stage ranges from 8 to 10 days or even beyond two months depending upon the atmospheric temperature. The mean lifespan of female moth has been recorded to be 20 days at 20°C, 7 days at 30–32°C and 4 days at 40°C. The life cycle is completed in four weeks at higher temperature or may require even six months under lower temperature conditions.

The development period of *G. mellonella* varies with combs of honeybee species and was recorded faster on combs of *A. cerana* followed by *A. dorsata* and *A. florea* (Table 9.2). It successfully completes its life cycle on the combs of major species of honeybees but fails to develop on processed beeswax. The combs of *A. mellifera* had lesser fecundity with prolonged larval and pupal stages which could be attributed to the presence of higher content of propolis. However, the combs of *A. laboriosa* does not appear to be attacked by wax moth as these moths are unable to survive in the cool climate and freezing winter conditions (Underwood, 1986). The greater wax moth attacks more vigorously in the colonies of Africanized bees than those of European bees. The longevity of moths is greater on the combs of *A. dorsata* followed by *A. cerana*. However, the duration of pre- and post-oviposition periods are recorded higher on the combs of *A. mellifera* (Table 9.3).

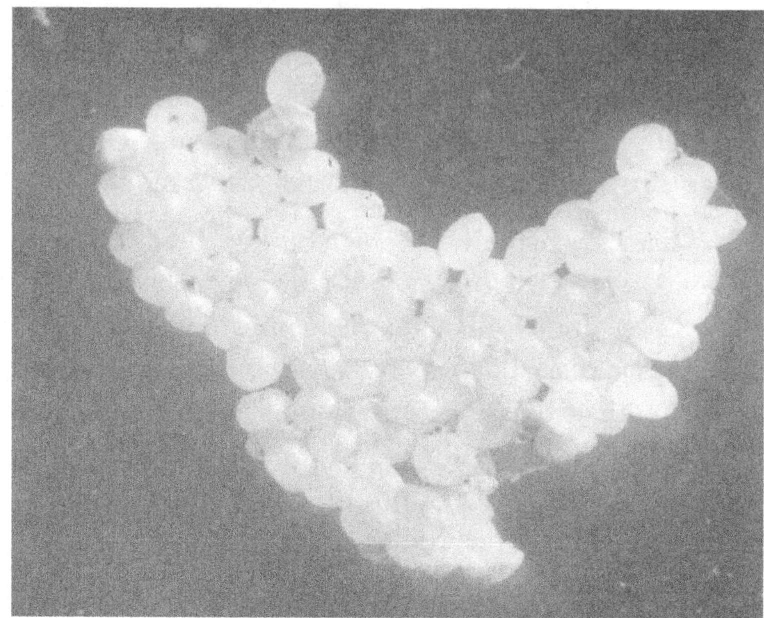

(a)

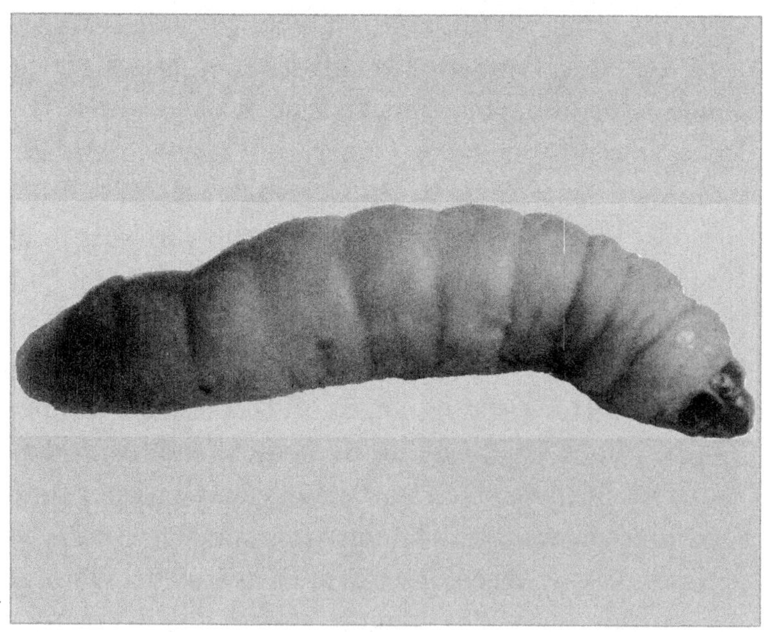

(b)

(c)

(d)

**Figure 9.3**   Developmental stages of greater wax moth, *Galleria mellonella* (a) Egg mass (b) Larva, (c) Pupa (d) Female moth

**Table 9.2**    Duration of different developmental stages of *Galleria mellonella* on combs of honeybee species (Hanumanthaswamy *et al.*, 2001)

| Honeybee species | Egg | Larva | Pre-pupa | Pupa | Total life period |
|---|---|---|---|---|---|
| *Apis cerana* | 8.60 ± 0.52 | 49.30 ± 1.66 | 2.10 ± 0.55 | 8.60 ± 0.75 | 68.60 ± 3.48 |
| *Apis dorsata* | 8.60 ± 0.52 | 54.20 ± 1.32 | 2.35 ± 0.49 | 9.40 ± 0.60 | 74.55 ± 2.93 |
| *Apis florea* | 8.70 ± 0.84 | 58.60 ± 3.03 | 2.40 ± 0.50 | 9.65 ± 0.75 | 79.35 ± 5.12 |
| *Apis mellifera* | 9.20 ± 0.63 | 110.65 ± 8.28 | 2.60 ± 0.68 | 10.05± 0.69 | 132.50 ±10.28 |

(Mean ± SD in days)

**Table 9.3**    Longevity and oviposition periods of *Galleria mellonella* on combs of honeybee species (Hanumanthaswamy, 2000)

| Honeybee species | Longevity (days) | | Oviposition period (days) | |
|---|---|---|---|---|
| | Male | Female | Pre-ovi position | Post-ovi position |
| *Apis cerana* | 16.40 ± 2.84 | 6.90 ± 0.74 | 1.10 ± 0.32 | 1.20 ± 0.42 |
| *Apis dorsata* | 17.50 ± 1.51 | 7.40 ± 1.26 | 1.10 ± 0.32 | 1.30 ± 0.48 |
| *Apis florea* | 16.00 ± 1.94 | 6.50 ± 1.18 | 1.20 ± 0.42 | 1.40 ± 0.52 |
| *Apis mellifera* | 15.70 ± 3.53 | 5.90 ± 0.88 | 1.30 ± 0.48 | 2.20 ± 1.23 |

(Mean ± SD in days)

## Management

***Manipulative methods***  In managing the wax moths special attention needs to be given to insecticide application, since the insecticide against the pest may affect the bee colonies. The best way to prevent wax moth menace is to keep bee colonies strong, hygienic and healthy with adequate food storage. Minimizing cracks and crevices in the hive, providing artificial feeding during dearth period and removal of unoccupied old combs are recommended in the apiary. The physical methods of control involve exposure of infested bee combs to higher or lower temperatures.

Exposing combs to a temperature of 55°C for a few minutes would kill all stages of wax moth (Naim and Bisht, 1972). The male moths may be sterilized with gamma rays routinely so that the mated females do not produce any offspring.

***Biological control***   Biological control of wax moths is possible with host-specific entomopathogenic viruses, bacteria and fungi. The multiple embedded nuclear polyhedrosis virus is sprayed in a suspension on empty infested combs (Dougherty *et al.*, 1982). Similarly, densonucleosis (Gross *et al.*, 1990) and nodamura viruses (Garzon *et al.*, 1978) have been reported to be effective against wax moth larvae.

Application of *Bacillus thuringiensis* (Bacillales: Bacillaceae) cause cent per cent mortality on various stages of wax moth larvae when fed through the diet (Arraras *et al.*, 1986). Lethal concentrations of many commercial products of *B. thuringiensis* such as baltospeine, certan, dipel and thuricide have been evaluated, of which certan showed higher toxicity followed by dipel, baltospeine and thuricide against wax moth larvae (Hanumanthaswamy, 2000).

Leephitakrat *et al.* (1999) evaluated the toxicity of three strains of *B. thuringiensis* on the larvae of both the species of wax moths. Among these *B. dendrolium* and *B. thuringiensis entomocidus* are highly toxic. Valdes (1974) reported that dipping of the wax moth larvae in a suspension of the fungus, *M. anisopliae*, recorded a mortality of 97 per cent in 10 days. Similarly, the strains (K87 and K2) of *B. bassiana* are also effective on larvae of wax moth (Pavlyushin, 1976).

The braconid wasp, *Apanteles galleriae* Wilkinson (Hymenoptera: Braconidae), is a major parasite found on the larvae of *G. mellonella*. The adult parasite is black in colour (Figure 9.4) and normally occurs on wax moth larvae with peak parasitization during September (Viraktamath *et al.*, 2005).

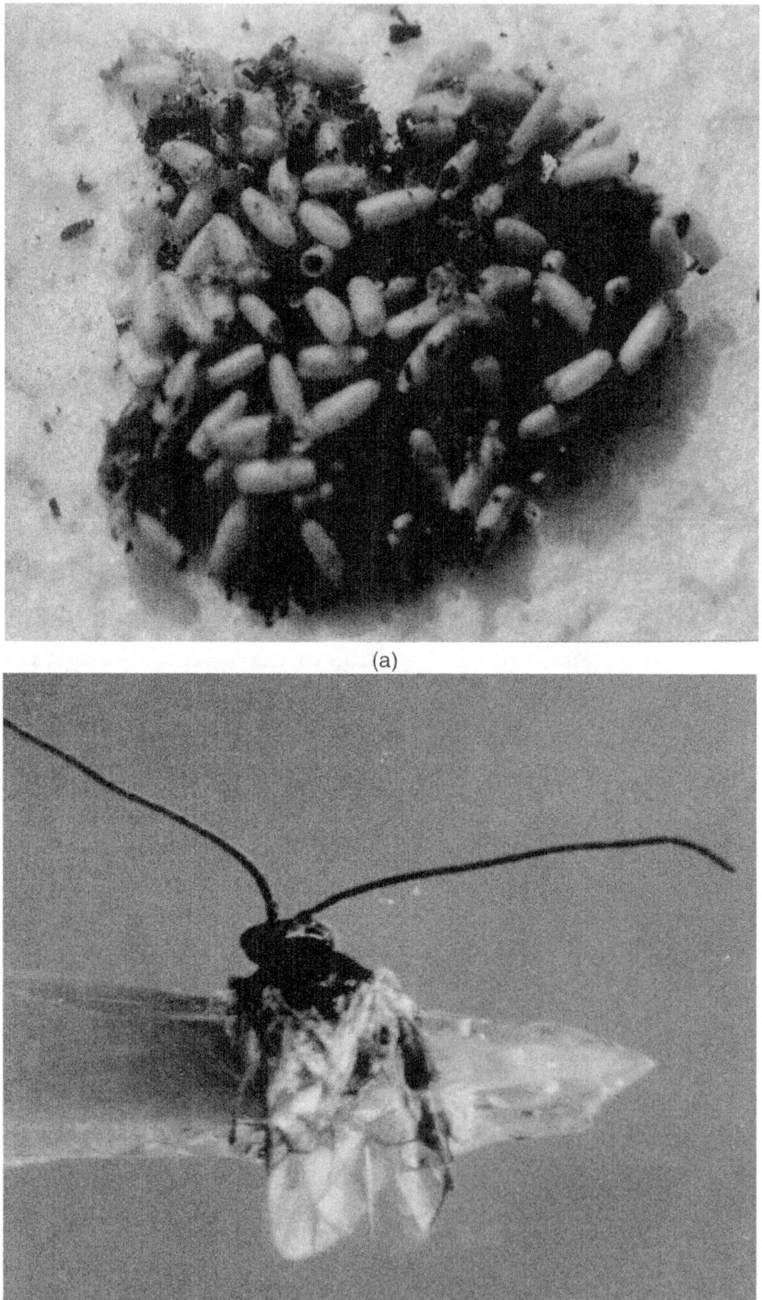

(a)

(b)

**Figure 9.4**   (a) Pupae and   (b) adult of parasitoid, *Apanteles galleriae*

The parasite lays few eggs on the wax moth larva and only one egg suceeds in completing the life cycle. The parasitic larvae emerge out of the wax moth larva by rupturing, and pupate in a small white silken cocoon. The parasitic wasps, *Trichogramma* spp. (Hymenoptera: Trichogrammatidae), parasitize on the eggs of wax moth. Releasing of these wasps at an interval of three weeks for 3 to 4 months ensure better parasitization.

***Chemical control***   Chemicals can be applied on combs to kill developmental stages of moths under storage conditions. These involve fumigation of combs with volatile compounds. Ethylene dibromide (EDB), paradichlorobenzene (PDB) and carbon dioxide ($CO_2$) are the commonly used fumigants. Paradichlorobenzene is one of the least hazardous and used to kill the larvae, pupae and adult wax moths but it is not effective against eggs. During fumigation, the hive parts are stacked tightly and cracks and crevices are sealed with adhesive strips. Two to three tablespoons of the crystals may be placed on the frames of the top super with a piece of paper. The chemical crystals may also be sprinkled directly on the top bars of the frames (Kapil and Sihag, 1983).

Carbon disulphide is highly inflammable and its vapours are highly explosive on mixing with air. It is used preferably in a well-ventilated shed and is effective against larvae and adult moths but not on eggs. The fumes from sulphur burning also would kill the larvae and adult wax moths.

Chlorosol, a mixture of methyl bromide and carbon tetrachloride is a strong fumigant that kills all the stages of wax moths including eggs. Spraying with a solution of the hydrochloride on brood combs prevents the development of wax moths (Feldlaufer *et al.*, 1998). The empty combs in the hive may be removed and stored after fumigation with ethylene dibromide in tightly enclosed containers. The sterol inhibitor, dimethyl dodecylamine, prevents the conversion of plant sterols into cholesterol in most insects.

## *Achroia grisella* Fab.

The lesser wax moth is not so serious pest compared to the greater wax moth in honeybee colonies. The life history of the former is more or less similar to the latter. Females lay eggs in cracks and crevices of the hive parts and also on the combs. The freshly hatched larvae burrow into the midrib of the combs. They feed and develop into adult in one to six months depending on the ambient temperature. The larva reaches its full size, starts to spin a silken cocoon and undergoes pupation. The cocoons are papery in texture, strong and normally white to brown in colour. The pupation occurs on the comb or in the loose debris on the bottom board of the hive. Males do not rely solely on pheromones to find a mate but also use ultrasound. Adult females are 10 to 15 mm long and wings are folded at a shallow angle over the top of the body. The wings bear scales and are grey to brown with a minor bronze tinge. The peak infestation of lesser wax moth could be seen during dearth period to the honeybees (Figure 9.5).

Another species of lesser wax moth, *Achroia innotata,* has been recorded in feral nests of *A. cerana* from Japan (Okada, 1988). The management practices that are used against greater wax moth may also be effective on lesser wax moths. The ultrasonic sound is effective in moth attraction especially in lesser wax moth. Spangler (1984) developed a stimulator of male making sound, which attract females and such devices could be used to detect and monitor wax moth infestation in honeybee colonies.

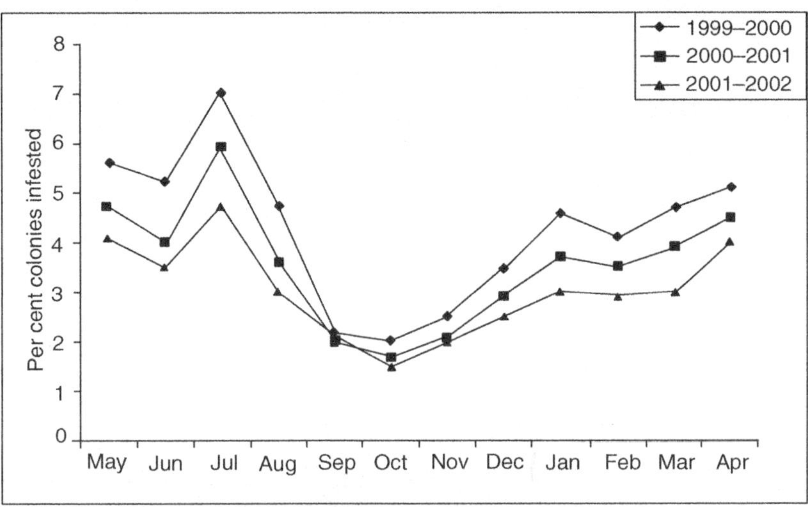


**Figure 9.5**    Infestation of lesser wax moth, *Achroia grisella* in the colonies of *Apis mellifera*

## HIVE BEETLES

### *Aethina tumida* Murray

The small hive beetle, *Aethina tumida*, is a destructive pest of honeybee colonies, causing damage to combs containing stored honey and pollen. It is widely distributed throughout the tropical and sub-tropical regions of Africa, USA and Australia (Hood, 2004). The female beetle lays irregular masses of eggs in cracks and crevices of the hive. The eggs hatch into larvae in about 2 to 3 days which feed on the pollen and honey by damaging the combs. The larvae are elongated, whitish and the grown-up larvae leave the hive and burrow into the soil near the hive for pupation. The pupal period may last for 3 to 4 weeks. The adult beetles are broad with a flattened body about 5 to 7 mm long and dark brown to black in colour (Figure 9.6). They complete 4 to 5 generations in a year during favourable conditions.

(a)

(b)

**Figure 9.6** (a) Larva and (b) adult of small hive beetle, *Aethina tumida*

## Nature of Damage

It is an endemic scavenger in the colonies of *A. mellifera* inhabiting sub Saharan Africa and occasionally seen damaging honeybee colonies. The damage is usually restricted to weak and diseased colonies. The pest feeds on pollen, honey and brood in its native range and has proven deleterious to the colonies of *A. mellifera* causing greater damage to the commercial beekeeping industries in its newly introduced regions of USA and Australia (Ellis and Hepburn, 2006). The rapid spread and high reproductive ability of the pest together with its capacity to hibernate in honeybee colonies have made it a potential pest worldwide (Neumann and Elzen, 2004). The females create holes in the wax capping of brood cells and deposit eggs on the pupae but the honeybees remove the oviposited brood. The beetles are attracted to nearby colonies and are suspected to transmit certain bee diseases.

## Management

***Manipulative methods*** Reducing the height of the hive entrance and placing metal screens prevent the inward entry of the beetles into the hives. Several beetle traps have been designed and most of them rely on collection of adult beetles into a reservoir of vegetable oil through slots in a horizontal screen/shallow box lid. Elzen *et al.* (1999a) developed a plastic bucket trap with hardware cloth mesh for collection of adult beetles. This trap is baited with a combination of hive products like honey, pollen, brood and adult bees and are found to be most attractive. A trap using a piece of cardboard with one side stripped off to expose the corrugations attract beetle larvae and adults when placed on hive floor (Elzen *et al.*, 1999b). Hood and Miller (2003) evaluated an in-hive trapping device supplied with attractants like alcohol, beer, ethylene, mineral oil, honey and cider vinegar. Among these, the cider vinegar containing trap yielded the maximum number of dead beetles in the field.

***Biological control*** The entomopathogenic nematodes, *Heterorhabditis megidis*, Poinar (Rhabditida: Heterorhabditidae) and *Steinernema carpocapsae* Weiser (Rhabditida: Steinernematidae) are known to infest pre-pupae of *A. tumida* (Cabanillas and Elzen, 2006). Similarly, susceptibility of adult beetles to the fungi, *M. anisopliae, B. bassiana* and *H. illustris* have been recorded (Ellis *et al.*, 2004, Muerrle *et al.*, 2006).

***Tolerance against beetles***   Hive beetles are generally found in spaces between the frames and bars of the hive. They move straight into a cluster of bees and sometimes they are decapitated by worker bees. However, they usually stay motionless and tuck their head underneath the pronotum with the legs and antennae pressed tightly into the body. The guard bees try to attack the beetles which move to the edges and keep them imprisoned. Bees add propolis on the edges of the hive around detected hidden beetles and encapsulate them. This confinement behaviour followed by aggression towards free roaming beetles and removal of eggs oviposited on capped brood cells is effective in regulating the beetle menace in honeybee colonies (Ellis *et al.*, 2003).

***Chemical control***   The ground around the apiary may be drenched with a synthetic pyrethroid, permethrin. The in-hive application of coumaphos-impregnated plastic strips on the bottom board of a colony kills the moving beetles (Elzen *et al.*, 1999b). Morse and Flottum (1997) recommended the use of paradichlorobenzene to fumigate the stored combs infested with the larvae of *A. tumida*.

## OTHER HIVE BEETLES

The large hive beetle, *Oplostomus fuligineus* (Coleoptera: Scarabaeidae) feeds on the honey. It is about 20 mm long with a strong integument. Other uncommon hive beetles recorded are *Diplognatha gagates* Forster, *Coenochilus* spp. (Coleoptera: Cetoniidae) and *Rhizoplatys trituberculatus* Burmeister (Coleoptera: Dynastidae) in Africa and *Platybolium* spp. (Coleoptera: Tenebrionidae) and *Dermestes vulpinus* Fab. (Coleoptera: Dermestidae) in tropical Asia.

# BEE LOUSE

## *Braula coeca* Nitzsch

*Braula coeca* (Diptera: Braulidae) commonly known as bee louse is a wingless dipteran. It is an oviparous insect, laying eggs on the wax capping. The eggs are also deposited on empty cells and on the bottom board but the eggs laid on the honey capping hatch successfully.

The larvae construct a tunnel under the capping of the combs and feed upon honey and pollen. The tunneling causes damage to the honey combs. The development period of the louse is about 3 to 4 weeks. The adult lice attach to the bee's hair, usually in the regions of head and thorax and feed on mouthparts of bees (Imms, 1942). These are often referred to as harmless but cause nuisance to bees. The lice are found on the worker, drone, and queen honeybees (Smith 1978). Queen bee is likely to collect more lice and her egg-laying capacity may be reduced. These are known to spread through robber bees, drifting bees, bee swarms and also by beekeepers.

Sprinkling of honey water on infested colonies triggers the bees to remove lice while cleaning up the honey. Application of tobacco smoke also kills the bee lice population.

# ANTS

Ants (Hymenoptera: Formicidae) generally live in underground or on arboreal nests of plants or trees. These are destructive predators of honeybees and cause heavy loss to beekeeping in tropical and sub-tropical regions. The weaver ant, *Oecophylla smaragdina*, black ants, *Camponotus compressus*, *Camponotus rufoglaucus*, and the small brown ant, *Monomorium* spp. are major predators of bee colonies. The ant species which commonly predate on honeybee colonies are presented in Table 9.4.

**Table 9.4**   Common predatory ants associated with honeybee species

| Ant species | Honeybee species | Status |
|---|---|---|
| **Formicinae** | | |
| *Camponotus compressus* Fab. | *Apis mellifera, Apis cerana, Apis dorsata, Apis florea* | Major |
| *Camponotus rufoglaucus* Jerd. | *Apis mellifera, Apis cerana, Apis dorsata, Apis florea* | Major |
| *Oecophylla smaragdina* Fab. | *Apis cerana, Apis mellifera, Apis florea* | Major |
| *Anoplolepis longipes* Jerd. | *Apis mellifera* | Minor |
| *Paratrechina longicornis* Latr. | *Apis cerana* | Minor |
| *Solenopsis geminata* Fab. | *Apis cerana* | Minor |
| **Myrmicinae** | | |
| *Monomorium floricola* Jerd. | *Apis mellifera, Apis cerana* | Major |
| *Crematogaster* spp. | *Apis cerana* | Minor |
| *Pheidologeton diversus* Jerd. | *Apis cerana* | Minor |
| *Tetramorium* spp. | *Apis cerana, Apis mellifera* | Minor |

*Oecophylla smaragdina* is a serious pest of bee colonies. It constructs the nest in bushy plants and trees by weaving the leaves with sticky threads in the form of a ball-like structure. These ants attack the bee colonies and eat or carry away the comb contents such as honey, pollen, brood and adult bees (Figure 9.7). *Apis florea* has developed a special defensive mechanism against the ants. It protects its nests with two rings of sticky bands around the substratum, one on each side of the comb. These sticky bands are roughened through their secretions which act as repellent against ants (Lindauer, 1957). Sometimes when the ants attack the bee colony, the worker bees counter attack the intruders. Suppose the intruder lands on the comb, they are attacked by bees clumping together (Pirk *et al.*, 2002). The ants generally forage in groups and

(a)

(b)

**Figure 9.7**     (a) Nest of weaver ant   (b) Predation of ants on the pupa of honeybee

kill the weaker bee colonies. Greater ant predation is recorded during rainy and winter seasons and may be attributed to the availability of abundant ant nests in and around the apiaries (Figure 9.8). The ants initially get attracted towards dead bees near the hive and later develop a tendency to kill the bee colonies (Nagaraja and Rajagopal, 2003b).

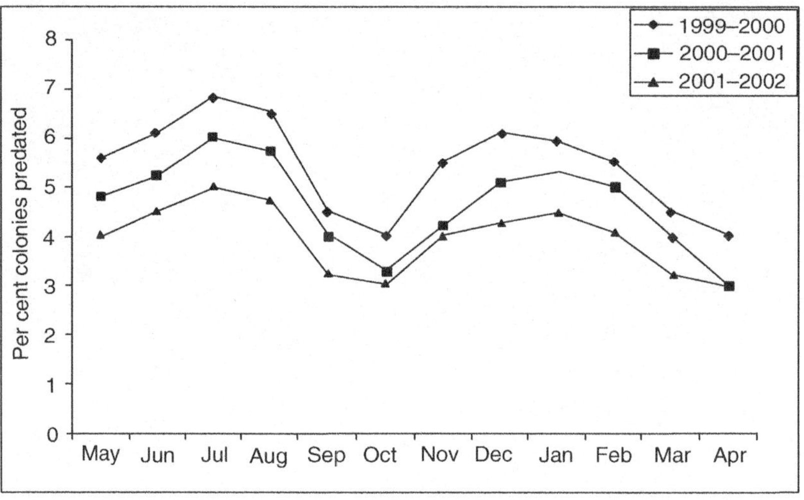


**Figure 9.8**  Predation of ants in the colonies of *Apis mellifera*

## Management

Ants can be prevented into beehives by providing ant pans filled with water around the base of the stand or smearing of oil bands over the stand. The stands could also be applied with gelatinous substances soaked in corrosive sublimate or engine oil, which are effective but last only for a week (Yusof and Ibrahim, 1995). The ant nests can be destroyed in and around the apiaries by treating with carbon disulphide, by punching holes in the earth around the nests. Direct application of lindane (0.2 per cent) or chlorpyrifos (0.1 per cent) is also effective in killing the colonies of ants. Poison baits may be placed near the ant nests in tin containers with openings too small for bees to enter.

# WASPS

Wasps (Hymenoptera) are widely distributed predators of honeybees and cause greater menace in many parts of the Asian continent. The food of wasps generally consists of the brood and adults of various insects and sweet meats. They attack bees at the hive entrance and also while foraging on the flowers. The common predatory wasps associated with honeybee species are presented in Table 9.5. *Vespa* spp. construct papery nests in hollow spaces of trees, corners of buildings and in bushes. (Figure 9.9a)They catch the bees, macerate and feed to their young ones. The wasp which is causing greater damage to bee colonies is *Vespa mandarina* commonly called as giant hornet in the Middle East and Asia. It constructs nest in the hollows of tree trunks and attacks the bee colonies en masse by taking over both brood and adult bees to feed larvae in their nests.

*Vespa tropica* (yellow-banded wasp) is an efficient flier and catches foraging bees (Figure 9.9b). These wasps which build nests in the ground are observed to fly around the beehives and catch bees at the hive entrance and at the field. Its predation has been observed throughout the year with peak from June to September that often coincides with the floral dearth period (Hanumanthaswamy, 2000). *V. orientalis* is a large and deep-brownish wasp with yellow bands across the abdomen. It builds nests generally in hidden places in the walls and hollows of trees. It is very common on the sweet meat shops and occasionally predate on honeybees.

The brown wasp, *Vespa velutina,* is a notorious predator of honeybees and is rust-red in colour and is hairy. It makes papery nests on the top of trees and catches foraging bees while returning to the hive. These are troublesome for beekeeping in hilly regions and are active from July to September (Shah and Shah, 1991). These wasps predate on

(a)

(b)

**Figure 9.9** (a) Nest of predatory wasp  (b) Adult wasp of *Vespa tropica*

**Table 9.5**     Common predatory wasps associated with honeybee species

| Wasp species | Honeybee species | Reference |
|---|---|---|
| Vespidae | | |
| *Vespa mandarina* Smith | *Apis cerana* | Ono *et al.* (1995) |
| *Vespa tropica* Linn. | *Apis mellifera* | Srivastava *et al.*(1995) |
| | *Apis cerana* | Subbaiah and Mahadevan (1958) |
| *Vespa orientalis* Linn. | *Apis mellifera* | Muzaffar and Ahmed (1986) |
| | *Apis cerana* | Kshirsagar and Mahindre (1975) |
| *Vespa velutina* Lepel | *Apis cerana* | Kshirsagar and Mahindre (1975) |
| *Vespa basalis* Smith | *Apis cerana* | Srivastava *et al.*(1995) |
| *Vespa cincta* Fab. | *Apis cerana* | Subbaiah and Mahadevan (1958) |
| *Vespa magnifica* Smith | *Apis mellifera* | Rahman and Rahman (1995) |
| *Vespula vulgaris* Linn. | *Apis mellifera* | Chang *et al.*(1993) |
| *Polistes gallicus* Linn. | *Apis mellifera* | Manino and Patetta (1987) |
| *Stelopolybia pallipes* Olivier | *Apis mellifera* | Machado *et al.* (1987) |
| Sphecidae | | |
| *Philanthus triangulum* Fab. | *Apis cerana* | Thomas and Thomas (1972) |
| *Philanthus ramakrishnae* Turner | *Apis cerana* | Singh (1962) |
| *Palarus latifrons* | *Apis mellifera* | Anderson *et al.* (1983) |
| *Palarus orientalis* Kohl | *Apis cerana* | Cherian and Mahadevan (1937) |

bee colonies throughout the year. The results obtained on the rate of predation by different species of wasps in Bangalore, India, showed greater predation in winter and low in early rainy seasons (Figure 9.10). The high predation is due to greater availability of foragers in the field and also at the hive entrance.

*Vespula* is distributed in temperate regions and is commonly known as yellow jacket. *Vespula vulgaris* causes severe damage to bee colonies. Among the bee wolves,

*Philanthus triangulum,* has been recorded on *A. mellifera.* *Philanthus ramakrishnae* preys on the hill variety of *A. cerana* in India (Singh, 1962). The female wasp normally catches the returning foragers near the hive entrance. It paralyses the prey by the process of stinging and later either takes away to the nest or squeezes the nectar out of the honey sac and imbibes by leaving the empty bee on the ground.

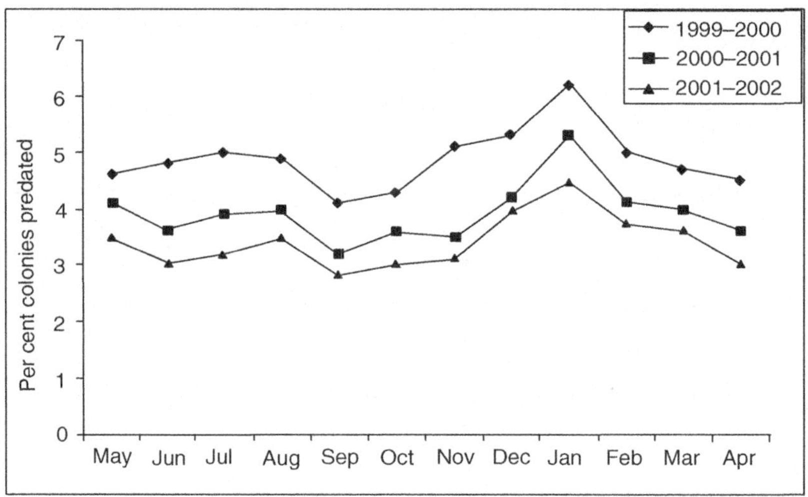


**Figure 9.10**  Predation of wasps in the colonies of *Apis mellifera*

The banded bee pirate, *Palarus latifrons,* is considered to be the most destructive honeybee predator (Anderson *et al.,* 1983). It nests gregariously in sandy soil and usually catches returning foragers near the hive entrance which later paralyse and squeeze the nectar from them. *Palarus orientalis* has been recorded as a predator of honeybees in India (Smith, 1960).

## Management

***Manipulative and chemical methods***  The routine methods advocated to control wasps in apiaries is collection and killing them during active predation. Finding nests through wasp lining has been developed in some regions of

southeast Asia. A foraging hornet is captured alive and a string is tied around its thorax and released. When the hornet flies off towards its nest, the wasp nest could be identified using the string as cue and the entire colony may be killed. Destruction of wasp nests with kerosene torches, fumigation with calcium cyanide or aluminum phosphide and spraying carbaryl on the nests are being practised. Feeding food impregnated with an insecticide through foraging hornets kills the individuals in their nests. The poisoned jaggery packed in gelatin capsules are glued to the thorax of the trapped foragers. On reaching to the nests, the poisoned jaggery would be shared by the nest mates thereby killing the entire colony.

***Tolerance against wasps*** Generally during wasp predation, hundreds of *A. cerana* worker bees guard around the hive entrance. When the wasp approaches the bee colony, most of the defending bees perform a side-to-side body-shaking behaviour. This behaviour causes the wasp to temporarily retreat for a few centimetres. Immediately worker bees catch and form a tight ball around the wasp and raise the temperature to a lethal level by killing the wasp (Ono *et al.*, 1995). This behaviour is apparently also exhibited by *A. nuluensis* (Koeniger *et al.*, 1996). Further, body-shaking behaviour was frequently observed on the swarms of *A. cerana*. The bees on the surface of the swarm-cluster react with simultaneous body-shaking behaviour against landing of flying insects.

## OTHER INSECT PESTS

Mahadevan (1951) recorded an assasin bug, *Acanthaspis siva* Distant (Hemiptera: Reduviidae) predating on *A. cerana*. It pierces the body of the worker honeybee to suck the haemolymph. Rao *et al.* (1972) observed a preying mantis,

*Odontomantis micans* (Mantodea: Hymenopodidae), as a predator of honeybees. The spider, *Nephila kuhlii* Doleschall (Araneae: Theraphosidae) builds webs on shrubs during monsoon and autumn and traps the foraging bees. The death head's hawk moth, *Acherontia styx*, enters the hives during night time and sucks honey. Robber flies, leaf cutter bees and dragon flies capture the bees and prey upon them.

*Manitoba acosta* parasitizes the drone brood of *A. mellifera* (Jelinski and Wojtowski, 1984). The larvae of the viviparous fly, *Senotainia tricuspis* Meig (Diptera: Sarcophagidae) parasitizes adult bees. The female fly deposits a newly hatched larva on bees entering into the hive by piercing through its mandibles to cause apimyiasis. The larva feeds on the soft body parts on death of the bee. Adult bees of *A. mellifera* are parasitized by the larvae of some viviparous flies such as *Physocephala* spp. (Diptera: Conopidae) and *Rondanioestrus apivorus* Vill (Diptera: Tachinidae) in South Africa (Anderson *et al.*, 1983).

Termites (Isoptera: Termitidae) such as *Heterotermes tenuis* Hag, *Eutermes costalis* Holmgren and *Microcerotermes arboreus* Emerson are known to damage the wooden beehives. Similarly many species of *Odontotermes*, *Microtermes* and *Coptotermes* are known to construct an earthen sheeting around the apiary and damage domestic beehives in India (Figure 9.11).

*Stylops* (Strepsiptera: Mengeidae) is a viviparous parasitic insect whose females are wingless and live within the abdominal cavity of the host. The newly emerged larva of these parasitic insects cling on to the body of the honeybee. It completes its life cycle by feeding on the larval haemolymph. Stylops also parasitizes *A. cerana* in both hills and plains of north India (Adlakha and Sharma, 1976).

**Figure 9.11** Damage by termites on the nucleus box of honeybee colony

## NON-INSECT PESTS

### *Ellingsenius indicus* Chamberlin

Pseudoscorpions occupy a selected niche depending on the microclimatic conditions. They prefer to live in different habitats with constant temperature. About a dozen species of pseudoscorpions have been reported to occur in bee hives. *Ellingsenius indicus* (Arachnida: Chelonethi) is commonly known to feed on micro-arthropod fauna associated with the nests of *A. cerana*.

The role of pseudoscorpions in honeybee colonies has not been well-documented although they are commonly associated with bee colonies of *A. cerana* and *A. mellifera*. Both nymphs and adults of pseudoscorpions are known to cling on to workers, drones and queen bees and they might prey upon bee parasitic mites. Since they are considered as minor associates of honeybees, no control measures have been advocated.

# 10

# Vertebrate Predators

Honeybee colonies are affected by a variety of vertebrate predators. They are attacked by amphibians, reptiles, birds and mammals. These predators cause severe damage to bee colonies and the extent of loss may vary with geographical regions.

## Toads and Lizards

The frogs and toads are known to prey upon varieties of insects and occasionally feed on the bees at the hive entrance. These are more proficient in capturing bees and are less affected by the bee stings and bee venom (Lescure, 1966). The giant toads, *Bufo melanostictus* Niger and *B. marinus* Linn. (Amphibia: Bufonidae) are the common amphibians recorded, which feed on the honeybees in different parts of the world.

The common southwestern lizard, *Sceloporus magister* Hallowell (Reptilia: Phrynosomatidae), also known as the twin-spotted spiny lizard or desert spiny lizard feeds on a wide variety of prey but they are not specific to honeybees.

It feeds on ants, beetles, spiders and small lizards (Degenhardt *et al.*, 1996). Geckos, *Hemidactylus brooki* Gray (Reptilia: Geckonidae), is a common predator of honeybees in different parts of India.

## MANAGEMENT

Placing bee colonies on hive stands prevent the entry of toads and lizards into the hive. Use of beehives free from cracks and crevices and also maintaining colonies with hygienic conditions would prevent the lizard problem.

## BIRDS

Birds are the major predators of honeybees. About 40 different species of birds are known to prey on bees in the field and also at the hive entrance. The beak of the birds are adapted to catch bees easily during flight. They manipulate the prey, dislodge the sting and remove the poison sac of the bees (Ambrose, 1990). The common birds which are known to predate on bee colonies are presented in Table 10.1.

The common green bee-eater, *Merops orientalis* preys upon a wide variety of insects in the Indian subcontinent. These birds are grass-green coloured with tinged reddish brown patches on the head with slender neck and slightly curved bill (Figure 10.1). These cause serious problem to honeybees in the plains, from May to August when they do not find other insects to prey upon. The bees are smapped up in the bill, and when the bird returns to its perch, beats up the prey until it dies. The degree of damage of bee colonies caused by birds depend largely upon their density. These birds seem to prefer queens and drones as they are visible easily. Bird predation in *A. mellifera ligustica* colonies was attributed to a greater extent to cloudy weather in the early rainy season (Figure 10.2). The blue-cheeked bird, *M. persicus,* and

rainbow bird, *M. ornatus,* are known to prey upon honeybees in Africa and Australia, respectively (Goebel, 1984). The blue-bearded bee-eater, *Nyctyornis athertoni,* is a predator of *A. dorsata.* It stimulates the bee's nest by making a quick fly fast and disturbing the nest. It then picks up the bees from its plumage and feeds upon them (Kastberger and Sharma, 2000).

**Table 10.1** Common bird predators of honeybee colonies

| Order/family | Common name | Scientific name | Status |
| --- | --- | --- | --- |
| Passeriformes | | | |
| Meropidae | Green bee-eater | *Merops orientalis* Latham | Major |
| | Rainbow bird | *Merops ornatus* Latham | Major |
| | Blue-bearded bee-eater | *Nyctyornis athertoni* Jardine & Selby | Major |
| Dicruridae | Black drongo | *Dicrurus adsimilis* Bechstein | Major |
| Corvidae | Common crow | *Corvus splendens* Vieillot | Minor |
| Accipitridae | Honey buzzard | *Pernis apivorus* Linn. | Minor |
| | Oriental honey buzzard | *Pernis ptilorhyncus* Temminck | Minor |
| Apodidae | Needle-tailed swift | *Hirundapus celebensis* Sclater | Minor |
| Picidae | Green woodpecker | *Picus viridis* Linn. | Major |
| Indicatoridae | Honey guide | *Indicator indicator* Vieillot | Minor |

The drongo, *Dicrurus adsimilis,* is commonly seen in rural areas with cultivated crops, perched on fence posts, bush-tops, telegraph wires, etc. (Figure 10.1). Apart from the insects, drongos eat smaller birds and sometimes taste flower nectar. The bird is found all over the Indian subcontinent, Bangladesh, Myanmar, Sri Lanka and Pakistan. These birds visit apiaries occasionally on cloudy days, and

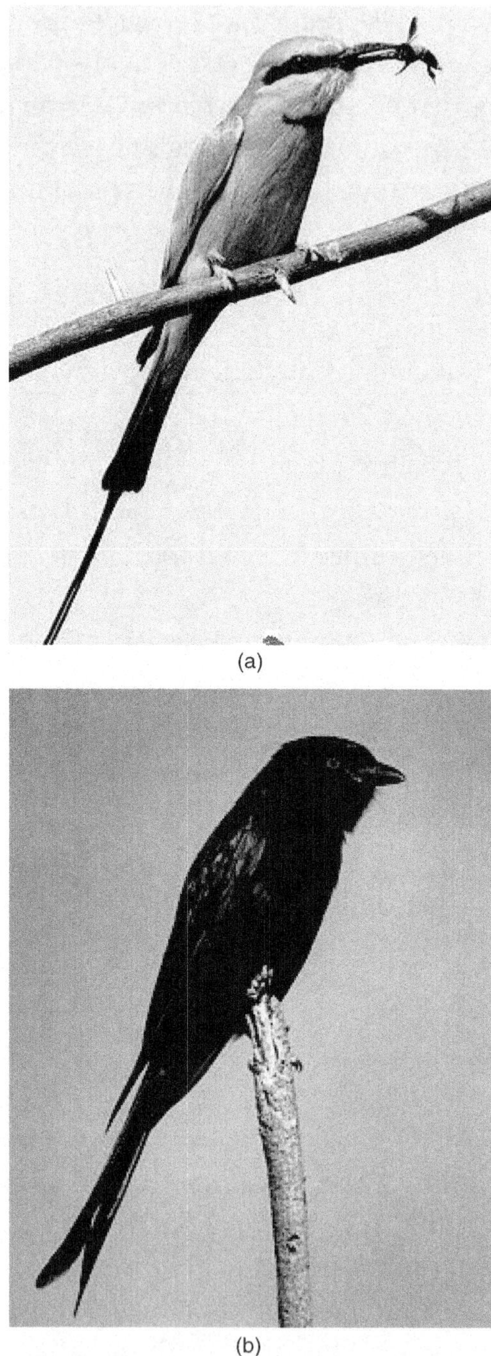

(a)

(b)

**Figure 10.1** a) Green bee-eater, *Merops orientalis* b) Drongo, *Dicrurus adsimilis*

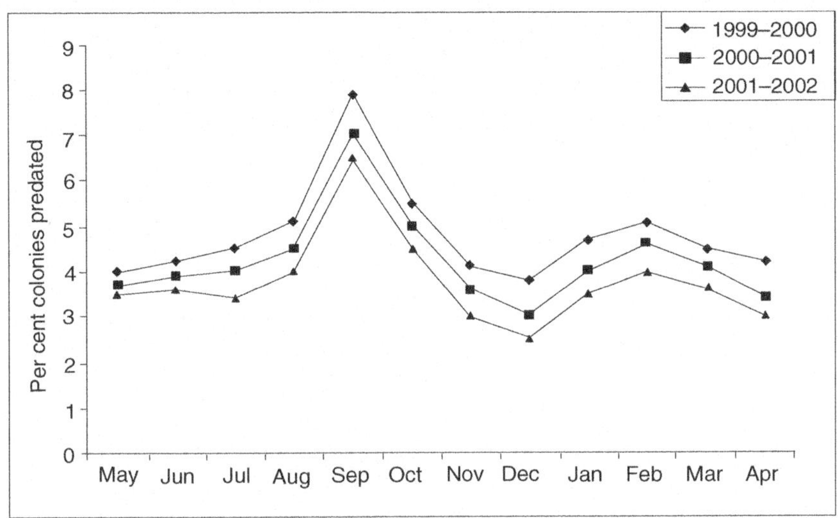


**Figure 10.2**   Predation of birds in the colonies of *Apis mellifera*

prey upon the bees. Studies at Bangalore, India, show that these birds capture bees throughout the day and their abundance is high in the morning hours. (Table 10.2). The common crow, *Corvus splendens* is a widely distributed predator of honeybees in India.

The face of the honey-buzzard, *Pernis apivorus*, is normally covered with small scalelike feathers, which are impregnable to bee stings. It mainly predates on wasps and bumble bees for its food but occasionally attacks honeybee nests. The oriental honey-buzzard, *Pernis ptilorhyncus*, is found throughout Asia and generally hunts *A. dorsata* colonies (Thapa and Wongsiri, 2003). Swifts prey on honeybees in flight and easily catch drone bees. The Philippine purple needle-tailed swift, *Hirundapus celebensis*, is one of the major constraints to beekeeping in the Philippines.

Woodpeckers have a strong, sharp, pointed bill for excavating insect brood holes in trees and a very long sticky tongue for extracting prey. They feed on a variety of insects and visit beehives when their regular food sources have become scarce. The green woodpecker, *Picus viridis,* has been reported from Europe to cause considerable damage by drilling holes through hive walls to catch the bees inside.

**Table 10.2**   Relationship between abundance of drongos in capturing of bees (Hanumanthaswamy, 2000).

| Time (Hrs) | Abundance of drongos | Number of drongos engaged in capturing | Number of bees captured |
|---|---|---|---|
| 06. 00 | 24.12 ± 1.58 | 7.50 ± 2.44 | 10.75 ± 3.76 |
| 07.00 | 31.37 ± 4.25 | 16.25 ± 4.08 | 27.87 ± 6.46 |
| 08.00 | 20.00 ± 6.48 | 14.62 ± 3.42 | 16.87 ± 3.58 |
| 09.00 | 16.00 ± 3.35 | 14.62 ± 3.42 | 16.87 ± 3.58 |
| 10.00 | 21.75 ± 7.74 | 20.05 ± 6.81 | 47.12 ± 20.50 |
| 11.00 | 16.87 ± 5.57 | 15.75 ± 5.11 | 25.50 ± 3.80 |
| 12.00 | 13.25 ± 2.10 | 12.62 ± 2.11 | 21.12 ± 6.95 |
| 13.00 | 11.12 ± 4.16 | 9.87 ± 2.89 | 12.87 ± 2.02 |
| 14.00 | 11.00 ± 4.03 | 9.25 ± 12.37 | 19.12 ± 30.74 |
| 15.00 | 15.00 ± 5.54 | 14.00 ± 5.22 | 25.75 ± 16.89 |
| 16.00 | 18.00 ± 4.92 | 15.50 ± 4.89 | 33.87 ± 22.12 |
| 17.00 | 11.00 ± 3.60 | 9.62 ± 3.23 | 17.12 ± 6.43 |

(Mean + SD)

Honey guides are small predatory birds related to woodpeckers, which eat the contents of honeybee nests. They are capable of digesting beeswax in the presence of microorganisms and yeast in the gut. The greater honey guide, *Indicator indicator,* on locating the honeybee nest, accompanies a mammal by perching, calling and guiding. This behaviour continues until the bird reaches the nest. The bird is known to feed on the adult bees, brood combs, pollen and honey.

## Management

The methods such as scaring, producing distress voice at high volume, restricting the flight using reflective tapes, compact discs, etc. have been advocated to prevent the bird menace in and around apiaries. Covering apiaries with strong mesh would prevent the entry and attack by the birds. *A. florea* on disturbance emits a high-pitched wing buzz. It elicits a hissing sound caused by wing vibration which may serve to deter the bird predators (Sen Serma *et al.*, 2002).

## MAMMALS

Mammals are major enemies of honeybee colonies. Mice (Rodentia: Muridae) are known to invade bee colonies for shelter, and destroy the combs (Figure 10.3). They feed on bees and hive products such as honey and pollen. They also damage beekeeping equipment in storage.

**Figure 10.3**   Honeybee comb damaged by mice

The best known and widely distributed mammalian predators are the bears (Carnivora: Ursidae). About five species of bears cause severe damage to bee colonies in parts of Asia, Europe, USSR, Canada and North America. They are black bear, *Ursus americanus* Pallas, brown bear, *Ursus arctos* Linn., Indian bear, *Melursus ursinus* Shaw, Asiatic black bear, *Selenarctos thibetanus* G.Cuvier and Malayan sun bear, *Helarctos malayanus* Raffles. Bears usually dismantle the hives to feed on the honey, pollen, brood and adult bees (Jadczak, 1986). They tear the hives into pieces and carry off honey combs to escape from mass stinging of bees. They also knock down the colonies of *A. dorsata*. Damage by bears has been shown to coincide with peak honey flow seasons.

Skunks are best known for their ability to excrete a strong foul-smelling faeces. *Mephitis* spp. and *Spilogale* spp. (Carnivora: Mephitidae) are natural enemies of bees reported from North America. They produce scants frequently in and around the apiary. Raccoons (Carnivora: Procyonidae) are nocturnal mammals which use thumbs to open beehives and damage the combs. In addition, opossums (Carnivora: Didelphidae) temporarily occupy abandoned burrows and secured areas. They damage wood of the hive entrance by chewing, and feed on the brood (Caron, 1978).

The honey badger, *Mellivora capensis* Storr (Carnivora: Mustelidae), is a widespread enemy of bees in the south of the Sahara. It has powerful claws in the forelegs and carries honey combs away from the hive. The nocturnal Indian honey badger, *Mellivora indica* Kerr, attacks hives in different parts of Asia. The Himalayan yellow-throated honey marten, *Martes flavigula* Boddaert, is distributed from south China to Malaysia and attacks bee colonies.

Many primates eat honey and brood of the bee nests. Rhesus monkey, *Macaca mulatta* Zimmermann (Primates:

Cercopithecidae), has been reported damaging brood of *A. florea*. Baboons may break up beehives and carry away the combs for consumption. Seeley *et al.* (1982) observed a troop of macaque monkeys shake a branch of the tree holding an *A. florea* nest until bees deserted the comb, after which they made off with the brood. The monkeys remove the adult bees from the combs and feed on the honey and brood. The problem of monkeys is also very common in south India where troops of monkeys attack the colonies of *A. cerana* and *A. florea*. The monkeys in troops jump on to the beehive, open the top cover and shake it for the bees to fly away. Later, they carry away with super and brood combs. Chimpanzees (Primates: Pongidae) have been observed hunting honeycombs of *A. mellifera* colonies in Africa (Van Lawick Goodall, 1968).

## Management

The suitable technique advocated to manage mice is to prevent their entry into the hives by using mouse guards. The hive entrances can be reduced by using wood, and metallic wires. Traps are quite useful in reducing the mice population. Identification of the live burrows around the apiaries and subsequent fumigation with hydrogen cyanide or aluminum phosphide controls mice menace in apiaries. Poison baits are also used to kill the mice.

Constructing strong normal or solar fences around the apiary could prevent the bear menace. Similarly, fixing hive guards can help to prevent the entry of skunks into the bee colonies (Jadczak, 1986).

Scaring monkeys away from the apiaries using loud noise, keeping bee colonies in the apiaries protected with nature's solar fence and tying bee boxes with strong wire are other methods which could reduce menace by monkeys.

# Conclusions

Honeybees are affected by varieties of natural enemies ranging from viruses to mammals. Thai sac brood virus disease is catastrophic to beekeeping industry with *A. cerana* in Asian continent. Its frequent infection declined the honeybee population in a rapid way. This has lead to a drastic reduction in honey, beeswax and crop pollination. There is a possibility of occurrence of the disease in many Asian honeybee species such as *A. koschevnikovi*, *A. nigrocincta* and *A. nuluensis*. It may also spread to open-nesting honeybee species such as *A. andreniformis* and *A. laboriosa* from species like *A. florea* and *A. dorsata* where its virulence have been reported. This may lead to unprecedented array in beekeeping with Asian honeybees.

Similarly, sac brood and Kashmir bee viruses have created a greater havoc in beekeeping with *A. mellifera* in most parts of the world. However, the impact of most of the other bee viruses shall not be neglected as they persist and cause inapparent infections and disease outbreaks seasonally.

Though a few preliminary management practices are quite effective, suitable drugs which actively suppress viral disease could be investigated. Research may be undertaken on methods of rapid detection of the viruses by the beekeepers at the apiary level. The continuous process of natural selection favours the development of tolerance against viral diseases. Systematic screening and breeding of selective tolerant bee stock may be given prime importance to reduce the problems. In addition, transmission of viral diseases through vectors such as microsporidian and parasitic mites could be prevented.

The outbreak of American foul brood could be associated with successive survival of their resting spores for long periods. American foul brood and European foul brood diseases cause heavy damage to *A. mellifera* colonies almost worldwide. They are also known to infect few species of Asian domesticated and wild honeybee colonies. The bacterial diseases are possibly associated with colony stress and thus creating stress-free environment with adequate food source, and ample ventilation would be effective. The timely management practices with non-chemical methods and subsequent treatment with suitable antibiotics are effective in disease management. In addition, selective breeding programmes are being practised in developing disease-resistant stock. Such breeding practices may also be adopted against fungal diseases as they are predominant with *A. mellifera* in temperate regions.

The microsporidian parasite, *N. apis*, causes heavy colony loss by weakening the bee colonies. In addition, its role as a vector in the transmission of viral diseases such as black queen cell virus, bee virus Y and filamentous virus are manyfold higher than infection. Similarly, the protozoan parasite, *M. mellificae*, not only infects the adult bees, but

also transmits many diseases. Studies on the mode of transmission of pathogens through such vectors provides an insight into the management of most of the microbial diseases in honeybees.

The endoparasitic mite, *A. woodi,* renders a greater loss to beekeeping not only by causing acarine disease but also by transmitting many diseases from the infected to healthy colonies. *V. destructor* has become a global problem to beekeeping industry with *A. mellifera.* However, *V. jacobsoni* causes limited damage in the colonies of *A. cerana* in Asia. *T. clareae* affects *A. mellifera* colonies with greater loss in Asia. Similarly it may invade the colonies of *A. mellifera* in western countries similar to *V. destructor.* Though many acaricides have been developed, the tolerance of mites to most of the chemicals has led to unsuccessful management. The long-term measure required to reduce the mite problem is the multiplication of mite-tolerant stocks. Similarly, screening and multiplication of biocontrol agents is an alternative technique required for effective management of bee mites.

Wax moths cause a greater damage in tropical and sub-tropical regions of the world. These pests are responsible for higher rate of colony absconding in Asian honeybees. Maintaining bee colonies strong and hygienic especially during dearth seasons is extremely desirable. In addition, beekeepers are advised to take up suitable steps to prevent wax moth menace through timely management practices. The small hive beetle, which initiated its menace in South Africa, has been spread to many parts of the world. It seems inevitable that beetles may be able to switch over from *A. mellifera* to many Asian honeybee colonies, as the life cycle of these species is more or less similar. Therefore it is possible that if these beetles are introduced to Asia they may cause greater damage to *A. cerana* colonies than *A. mellifera.*

Ants and wasps are the major insect predators of honeybees in tropical conditions. These are known to attack huge number of bee colonies in spite of following general managment practices. There is a need to develop suitable management strategies against these predators in the apiaries as well as in the field. Similarly, the vertebrate predators such as different species of birds, bears and monkeys attack bee colonies almost throughout the world. Although few common management practices are available to prevent the attack of these predators, still suitable ecofriendly measures are required to be investigated. Research on these areas should be focused to safeguard the honeybee colonies from these animals.

In view of the above, suitable measures should be taken to prevent the spread of diseases and pests of honeybees by following strict quarantine measures while exporting or importing the bees. Improved technology especially on the application of suitable integrated management against various pests and diseases has to be recommended and such technology should be more sustainable, safer and easily accessible to the common beekeepers similar to the care and maintenance of the domestic animals.

# REFERENCES


Abrol, D.P. 2000. Beekeeping with *Apis cerana* in Jammu and Kashmir: present status and future perspectives. *Bee Wld.* 81: 149–152.

Abrol, D.P. and Putatunda, B.N. 1995. Discovery of an ectoparasitic mite *Tropilaelaps koenigerum* Delfinado-Baker and Baker (Acari: Laelapidae) associated with honeybees, *Apis dorsata, Apis mellifera* and *Apis cerana* in Jammu, India. *Curr. Sci.* 68: 90.

Adlakha, R.L. 1976. Acarine disease of adult honeybees in India. *Am. Bee J.* 116: 324–344.

Adlakha, R.L. and Sharma, O.P. 1976. *Stylops* (Strepsiptera) parasites of honeybees in India." *Am. Bee J.* 116: 66.

Aggarwal, K. 1990. Interaction between the parasitic mite, *Tropilaelaps clareae* (Acari: Laelapidae) in its host(s): Ecological and physiological studies. *Ph.D thesis*, Haryana Agricultural University, Hissar, India, p. 98.

Akratanakul, P. and Burgett, M. 1976. *Euvarroa sinhai* Delfinado and Baker (Acarina: Mesostigmata): a parasitic mite of *Apis florea* F. *J. Apic. Res.* 15: 11–13.

Alcock, J. 1993. *Animal behaviour: An Evolutionary Approach.* Sinauer Associates Inc, Massachusetts, USA. p. 525.

Alippi, A.M. 1994. Sensibilidad *in vitro* de *Bacillus larvae* frente a diferentes agentes antimicrobianos. *Vida Apicola* 66: 20–24.

Alippi, A.M., Ringuelet, J.A., Cerimete, E.L., Re, M.S. and Henning, C.P. 1996. Antimicrobial activity of some essential oils against *Paenibacillus larvae*, the causal agent of AFB disease. *J. Herbs Spices Med. Plants.* 4: 9–16.

Alizadeh, A. and Mossadegh, M.S. 1994. Stone brood and some other fungi associated with *Apis florea* in Iran. *J. Apic. Res.* **33**: 213–218.

Allen, M.F. 1995. Bees and beekeeping in Nepal. *Bee Wld.* **76**: 185–194.

Allen, M.F. and Ball, B.V. 1995. Characterization and serological relationships of strains of Kashmir bee virus. *Ann. Appl. Biol.* **126**: 471–484.

Allen, M.F. and Ball, B.V. 1996. The incidence and world distribution of honeybee viruses. *Bee Wld.* **77**: 141–162.

Allen, M.F., Ball, B.V., White, R.F. and Antoniw, J.F. 1986. The detection of acute paralysis virus in *Varroa jacobsoni* by the use of a simple indirect ELISA. *J. Apic. Res.* **25**: 100–105.

Ambrose, J.T. 1990. Birds: In: *Honeybee Pests, Predators and Diseases.* Morse, R.A. and Nowogrodzki, R. (eds.). Cornell University Press, Ithaca, USA. pp. 243–260.

Anderson, D.L.1984. A comparison of serological techniques for detecting and identifying honeybee viruses. *J. Inv. Pathol.* **44**: 233–243.

Anderson, D.L. 1985. Viruses of New Zealand honeybees. *New Zealand Beekeeper.* **188**: 8–10.

Anderson, D.L. 1993. Pathogens and queen bees. *Australian Beekeeper.* **94**: 292–296.

Anderson, D.L. 1994. Non-reproduction of *Varroa jacobsoni* in *Apis mellifera* colonies in Papua New Guinea and Indonesia. *Apidologie.* **25**: 412–421.

Anderson, D.L. 1995. Viruses of *Apis cerana* and *Apis mellifera*. In: *The Asiatic Hive bee: Apiculture, Biology and Role in Sustainable Development in Tropical and Subtropical Asia.* Kevan, P.G. (ed.). Enviroquest Ltd., Canada. pp.161–170.

Anderson, D.L. and Morgan, M.J. 2007. Genetic and morphological variations of bee parasitic *Tropilaelaps* mites (Acari: Laelapidae): new and re-classified species. *Exp. Appl. Acarol.* **43**: 1–24.

Anderson, D.L. and Trueman, J.W.H. 2000. *Varroa jacobsoni* (Acari: Varroidae) is more than one species. *Exp. Appl. Acarol.* **24**: 165–189.

Anderson, R.H. Buys, B. and Johannsmeier, M.F. 1983. Beekeeping in South Africa. *Bull.Tech. Serv.* No.394. Dept. of Agric., Pretoria, South Africa. p. 207.

Anderson, D.L. Giacon, H. and Gibson, N.L. 1997. Culture, detection and thermal destruction of the chalk brood fungus, *Ascosphaera apis*. *J. Apic. Res.* **36**: 163–168.

Arechavaleta-Velasco, M.E. Hunt, G.J. and Emore, C. 2003. Quantitative trait loci that influence the expression of guarding and stinging behaviours of individual honeybees. *Behav. Genet.* **33**: 357–364.

Arias, M.C. and Sheppard, W.S. 2005. Phylogenetic relationships of honeybees (Hymenoptera: Apinae) inferred from nuclear and mitochondrial DNA sequence data. *Mol. Phylogenetic. Evol.* **37**: 25–33.

Arias, M.S., Tingek, S., Kelitu, A. and Sheppard, W.S. 1996. *Apis nuluensis* Tingek, Koeniger and Koeniger and its genetic relationship with sympatric species inferred from DNA sequences. *Apidologie.* **27**: 415–422.

Aristotle (384–322 B.C.). *History of Animals, Book IX, 262.* Translated by Richard Crossware (1907), George Bell, London.

Arraras, E.A., Aracas, J.A. and Yantorno, O.M. 1986. Artificial rearing of *Galleria mellonella* and its use in measuring the bio-insecticidal activity of suspensions of *Bacillus thuringiensis* var. *kurstaki. Revista de la Facultad de Agronomia Universidad de Buenos Aires.* **7**: 71–76.

Atwal, A.S. 1971. Acarine disease problem of the Indian honeybee, *Apis indica* F. *Am. Bee J.* **111**: 134–135 and 186–187.

Atwal, A.S. and Goyal, N.P. 1971. Infestation of honeybee colonies with *Tropilaelaps* and its control. *J. Apic. Res.* **10**: 137–142.

Atwal, A.S. and Goyal, N.P. 1973. Introduction of *Apis mellifera* in Punjab plains. *Indian Bee J.* **35**: 1–9.

Atwal, A.S. and Sharma, O.P. 1970. Acarine disease of adult honeybees, prevention and control. *Indian Farming.* 20: 39–40.

Baggio, A., Gallina, A., Dainese, N., Manzinello, C., Mutinelli, F., Serra, G., Colombo, R., Carpana, E., Sabatini, A.G., Wallner, K., Piro, R. and Sangiorgi, E. 2005. Gamma radiation: A sanitation treatment of AFB contaminated beekeeping equipment: Gamma radiation sanitation in beekeeping management. *Apiacta.* 40: 22–27.

Bailey, L. 1966. The effect of acid-hydrolysed sucrose on honeybees. *J. Apic. Res.* 5: 127–136.

Bailey, L. 1969. The multiplication and spread of sac brood virus of bees. *Annl. Appl. Biol.* 63: 483–491.

Bailey, L. 1981. *Honeybee Pathology.* Academic Press, London. p. 124.

Bailey, L. and Ball, B.V. 1978. *Apis* iridescent virus and clustering disease of *Apis cerana. J. Inv. Pathol.* 31: 368–371.

Bailey, L. and Ball, B.V. 1991. *Honeybee Pathology.* Academic Press, London. p. 193.

Bailey, L. and Gibbs, A.J. 1964. Acute infection of bees with paralysis virus. *J. Insect Pathol.* 6: 395–407.

Bailey, L. and Woods, R.D. 1974. Three previously undescribed viruses from the honeybee. *J. Gen. Virol.* 25: 175–186.

Bailey, L. and Woods, R.D. 1977. Two more small RNA viruses from honeybees and further observations on sac brood and acute bee paralysis viruses. *J. Gen. Virol.* 37: 175–182.

Bailey, L., Ball, B.V. and Perry, J.N. 1983a. Honeybee paralysis: its natural spread and its diminished incidence in England and Wales. *J. Apic. Res.* 22: 191–195.

Bailey, L., Ball, B.V. and Perry, J.N. 1983b. Association of viruses with two protozoan pathogens of the honeybee. *Ann. Appl. Biol.* 103: 13–20.

Bailey, L., Carpenter, J.M. and Woods, R.D. 1979. Egypt bee virus and Australian isolates of Kashmir bee virus. *J. Gen. Virol.* 43: 641–647.

Bailey, L., Gibbs, A.J. and Woods, R.D. 1963. Two viruses from adult honeybees (*Apis mellifera* L.). *Virol.* **21**: 390–395.

Bailey, L., Ball, B.V., Carpenter, J.M. and Woods, R.D. 1980a. Small virus like particles in honeybees associated with chronic paralysis virus and with a previously undescribed disease. *J. Gen. Virol.* **46**: 149–155.

Bailey, L., Ball, B.V. Carpenter, J.M. and Woods, R.D. 1982. A strain of sac brood virus from *Apis cerana*. *J. Inv. Pathol.* **39**: 264–265.

Bailey, L., Carpenter, J.M., Govier, D.A. and Woods, R.D. 1980b. Bee virus Y. *J. Gen. Virol.* **51**: 405–407.

Ball, B.V. 1989. *Varroa jacobsoni* as a virus vector. In: *Present Status of Varroatosis in Europe and Progress in Varroa Mite Control*. Cavalloro, R. (ed.). Luxemburg. pp. 241–244.

Ball, B.V. and Allen, M.F. 1988. The prevalence of pathogens in honeybee (*Apis mellifera*) colonies infested with the parasitic mite, *Varroa jacobsoni*. *Annl. Appl. Biol.* **113**: 237–244.

Ball, B.V., Overton, H.A., Buck, K.W., Bailey, L. and Perry, J.N. 1985. Relationships between the multiplication of chronic bee paralysis virus and its associate particle. *J. Gen. Virol.* **66**: 1423–1429.

Batra, S.W. 1996. Biology of *Apis laboriosa*, a pollinator of apple at high altitudes in the greater Himalaya range of Garhwal, India (Hymenoptera: Apidae). *J. Kansas Entomol. Soc.* London. B **265**: 2009–2014.

Benjeddou, M., Leat, N., Allsopp, M. and Davison, S. 2001. Detection of acute paralysis virus and black cell virus from honeybees by reverse transcriptase PCR. *Appl. Environ. Microbiol.* **67**: 2384–2387.

Bennet, P.M. 1980. A review of the parasitic *Acarapis* mites of honeybees. *Australian Bee J.* **120**: 42–44.

Bhardwaj, R.K. 1968. A new record of the mite *Tropilaelaps clareae*; a parasite of honeybee in south east Asia. *Bee Wld.* **49**: 115.

Bienefeld, K. 1996. Factors affecting duration of the post capping period in brood of the honeybee (*Apis mellifera carnica*). *J. Apic. Res.* **35**: 11–17.

Bilikova, K., Wu, G. and Smith, J. 2001. Isolation of a peptide fraction from honeybee royal jelly as a potential antifoul brood factor. *Apidologie*. 32: 275–283.

Boecking, O. 1992. Removal behaviour of *Apis mellifera* colonies towards sealed brood cells infested with *Varroa jacobsoni:* techniques, extent and efficacy. *Apidologie*. 23: 371–373.

Boecking, O. and Drescher, W. 1992. The removal responses of *Apis mellifera* L. colonies to brood in wax and plastic cells after artificial and natural infestation with *Varroa jacobsoni* Oud. and to freeze killed brood. *Exp. Appl. Acarol.* 16: 321–329.

Boecking, O. and Spivak, M. 1999. Behavioural defenses of honeybees against *Varroa jacobsoni* Oud. *Apidologie*. 30: 141–158.

Bolli, H.K., Bogdanov, S., Imdorf, A. and Fluri, P. 1993. Initial results of the field treatment of honeybee colonies infested with *Varroa jacobsoni* using formic acid in hot climates. *Am. Bee J.* 129: 735–737.

Brar, H.S., Gatoria, G. S., Jhaj, H.S. and Chahal, B.S. 1985. Seasonal infestation of *Galleria mellonella* and population of *Vespa orientalis* in *Apis mellifera* apiaries in Punjab. *Indian J. Ecol.* 112: 109–112.

Breed, M.D. 1991. Defensive behaviour. In: *The African Honeybee.* Spivak, M., Fletcher, D.J.C. and Breed, M.D. (eds.) Boulder Co., Westview Press, USA. pp. 299–308.

Breed, M.D., Guzman-Novoa, E. and Hunt, G. J. 2004. Defensive behaviour of honeybees; organization, genetics and comparisons with other bees. *Ann. Rev. Entomol.* 49: 271–298.

Bromenshenk, J.J. 1979. Investigation of the impact of coal-fired power plant emissions upon insects. Rep. US Environ. Prot. Agency (EPA). pp. 215–239.

Bromenshenk, J.J., Carlson, S.R., Simpson, J.C. and Thomas, J.M. 1985. Pollution monitoring of puget sound with honeybees. *Science*. 227: 632–634.

Bruce, W.A., Anderson, D.L., Calderone, N.E. and Shimanuki, H. 1995. Survey for Kashmir bee virus in honeybee colonies in the United States. *Am. Bee J.* 135: 352–355.

Brückner, D.1975. Die Abhangigkeit der temperaturregulierung von der genetischen Variabilitat der Honigbiene (*Apis mellifera*). *Apidologie.* 6: 361–380.

Büchler, R. 1998. First results of a breeding programme on improvement of *Varroa* tolerance. *Pszczelnicze Zeszyty Naukowe.* 17: 15–16.

Büchler, R. and Drescher, W. 1990. Variance and heritability of the capped developmental stage in European *Apis mellifera* L. and its correlation with increased *Varroa jacobsoni* infestation. *J. Apic. Res.* 29: 172–176.

Büchler, R., Drescher, W. and Tornier, I. 1992. Grooming behaviour of *Apis cerana, Apis mellifera* and *Apis dorsata* and its effect on the parasitic mites, *Varroa jacobsoni and Tropilaelaps clareae. Exp. Appl. Acarol.* 16: 313–319.

Burges, H.D. 1978. Control of wax moths: physical, chemical and biological methods. *Bee Wld.* 59: 129–138.

Burgett, D.M. and Kitprasert, C. 1990. Evaluation of Apistan as a control for *Tropilaelaps clareae,* an Asian honeybee brood mite parasite. *Am. Bee J.* 130: 51–53.

Burnside, C.E. 1932. A newly discovered brood disease. *Am. Bee J.* 72: 433.

Butler, C.G. 1974. *The World of Honeybee.* Collins, London. p. 226.

Cabanillas, H.E. and Elzen, P. J. 2006. Infectivity of entomopathogenic nematodes (Steinernematidae and Heterorhabditidae) against the small hive beetle, *Aethina tumida* (Coleoptera: Nitidulidae). *J. Apic. Res.* 45: 49–50.

Calderone, N.W. and Kuenen, L.P. 2001. Effects of western honeybee (Hymenoptera: Apidae) colony, cell type and larval sex on host acquisition by female *Varroa destructor* ( Acari: Varroidae). *J. Econ. Entomol.* 94: 1022–1030.

Calesnich, E. J. and White, J.W. 1952. Thermal resistance of *Bacillus larvae* spores in honey. *J. Bacteriol.* **64**: 9–15.

Camazine, S. 1986. Differential reproduction of the mite, *Varroa jacobsoni* (Mesostigmata: Varroidae) of Africanized and European honeybees (Hymenoptera: Apidae). *Ann. Entomol. Soc. Am.* **79**: 801–803.

Caron, D.M. 1978. Marsupials and Mammals. In: *Honeybee Pests, Predators and Diseases.* Morse, R.A. (ed.). Cornell University Press, Ithaca and London. pp. 227–256.

Cassier, P., Tel-Zur, D. and Lensky, Y. 1994. The sting sheaths of honeybee workers (*Apis mellifera*): structure and alarm pheromone secretion. *J. Insect Physiol.* **40**: 23–32.

Chandler, D., Sunderland, K.D., Ball, B.V. and Davidson, G. 2001. Perspective biological control agents of *Varroa destructor* n sp., an important pest of European honeybee, *Apis mellifera. Biocontrol Sci. Technol.* **11**: 429–448.

Chang, Y.D., Lee, M.Y., Yim, Y.H. and Youn, Y.N. 1993. Species and visiting patterns of wasps (Hymenoptera: Vespidae) in an apiary. *Korean J. Apic.* **8**: 22–28.

Charbounneau, R., Gosselin, P. and Thibault, C. 1992. Irradiation of AFB. *Am. Bee J.***132**: 249–251.

Charriere, J.D. and Imdorf, A. 2002. Oxalic acid treatment by trickling against *Varroa destructor*: recommendations for use in central Europe and under temperate climatic conditions. *Bee Wld.* **83**: 51–60.

Chen, Y.P. and Siede, R. 2007. Honeybee viruses. In: *Advances in Virus Research,* Academic Press. pp. 33–80.

Chen, Y.P., Pettis, J.S., Collins, A. and Feldlaufer, M.F. 2006. Prevalance and transmission of honeybee viruses. *Appl. Environ. Microbiol.* **72**: 606–611.

Cherian, M.C. and Mahadevan, V. 1937. A new enemy of the Indian honeybee. *Madras Agric. J.* **25**: 35–37.

Cheshire, F.R. 1884. *Bees and Beekeeping: Scientific and Practical.* London. p. 336.

Chinh, P.H. 2000. Thai sac brood disease control in Vietnam. In: *Asian bee and Beekeeping Progress of Research and Development.* Matsuka, M., Verma, L.R., Wongsiri, S., Shrestha, K.H. and Uma Prathap. (eds.). Kathmandu, Nepal. pp. 57–59.

Chioveanu, G., Jonescu, D. and Mardare, A. 2004. Control of nosemosis: the treatment with Protofil. *Apiacta.* **39**: 31–38.

Clark, T.B. 1978. A filamentous virus of the honeybee. *J. Inv. Pathol.* **32**: 332–340.

Colin, M.E., Ducos, J., Larribau, E. and Boue, T. 1989. Activite des huiles essentielles de Labiees sur *Ascosphaera apis* et traitement dun rucher. *Apidologie.* **20**: 221–228.

Collins, A.M., Rinderer, T.E., Tucker, K.W., Sylvester, H.A. and Lackett, J.J. 1980. A model of honeybee defensive behaviour. *J. Apic. Res.* **19**: 224–231.

Cook, V.A. and Bowman, C.E. 1986. *Melitiphis alvearius*, a little known mite of the honeybee colony, found on New Zealand bees imported into England. *Bee Wld.* **64**: 62–64.

Crane, E. 1990. *Bees and Beekeeping: Science, Practice and World Resources.* Heinemann Newnes, Oxford. p. 614.

Dadant, C. 1890. *Langstroth on the Hive and Honeybee.* Dadant and Sons, Hamilton, Illinois, USA.

Dainat, B., Ken, T., Berthoud, H. and Neumann, P. 2009. The ectoparasitic mite, *Tropilaelaps mercedesae* (Acari: Laelapidae) as a vector of honeybee viruses. *Insect Soc.* **56**: (in press).

Dall, D.J. 1985. Inapparent infection of honeybee pupae by Kashmir and sac brood bee viruses in Australia. *Ann. Appl. Biol.* **106**: 461–468.

Dall, D.J. 1987. Multiplication of Kashmir bee virus in pupae of the honeybee, *Apis mellifera. J. Inv. Pathol.* **49**: 279–290.

Danka, R.G. 2001. Resistance of honeybees to tracheal mites. In: *Mites of the Honeybee.* Webster, T.C. and Delaplane, K.S. (eds.). Dadant and Sons, Hamilton, USA. pp. 117–129.

Davidson, E.W. 1973. Ultrastructure of American foul brood disease pathogenesis in larvae of the worker honeybee, *Apis mellifera. J. Inv. Pathol.* 21: 53–61.

Degenhardt, W.G., Painter, C.W., Price, A.H. and Garrett, C.M. 1996. *Amphibians and Reptiles of New Mexico.* University New Mexico Press, Albuquerque, USA.

de Guzman, L.I. and Delfinado-Baker, M. 1996. A new species of *Varroa* (Acari: Varroidae) associated with *Apis koschevnikovi* (Apidae: Hymenoptera) in Borneo. *Intern. J. Acarol.* 22: 23–27.

de Guzman, L.I. Rinderer, T.E. and Beaman, L.D. 1996. Attractiveness to infestation by tracheal mites, *Acarapis woodi* (Rennie) (Acari: Tarsonemidae) in three stocks of honeybees and two of their hybrids. *Bee Sci.* 4: 87–91.

de Guzman, L.I. Rinderer, T.E. and Delatte, G.T. 1998. Comparative resistance of four honeybee (Hymenoptera: Apidae) stocks to infestation by *Acarapis woodi* (Acari: Tarsonemidae). *J. Econ. Entomol.* 91: 1078–1083.

De Jong, D. 1997. Mites: *Varroa* and other parasites of brood. In: *Honeybee Pests, Predators, and Diseases.* Morse, R.M. and Flottum, P.K. (eds.). Medina, USA. pp. 281–327.

De Jong, D., Morse, R.A. and Eickwort, G.C. 1982. Mite pests of honeybees. *Ann. Rev. Entomol.* 27: 229–252.

de Ruiter, A. and vander Steen, J. 1989. Disinfection of combs by means of acetic acid (96%) against *Nosema. Apidologie.* 20: 503–506.

Delfinado, M.D. and Baker, E.W. 1961. *Tropilaelaps:* a new genus of mite from the Philippines (Laelapidae: Acarina). *Fieldiana Zool.* 44: 53–56.

Delfinado, M.D. and Baker, E.W. 1974. A new record for the bee mite, *Melitiphis* in New Zealand. *Bee Wld.* 55: 148–149.

Delfinado-Baker, M. 1982. New records for *Tropilaelaps clareae* from colonies of *Apis cerana indica. Am. Bee J.* 122: 382.

Delfinado-Baker, M. 1984. The nymphal stages and male *Varroa jacobsoni* Oudemans a parasite of honeybees. *Intern. J. Acarol.* 10: 75–80.

Delfinado-Baker, M. and Aggarwal, K. 1987. A new *Varroa* (Acari: Varroidae) from the nest of *Apis cerana* (Apidae). *Intern. J. Acarol.* 13: 233–237.

Delfinado-Baker, M. and Baker, E.W. 1982. A new species of *Tropilaelaps* parasitic on honeybees. *Am. Bee J.* 122: 416–417.

Delfinado-Baker, M. and Peng, C.Y.S. 1995. *Varroa jacobsoni* and *Tropilaelaps clareae:* a perspective of life history and why Asia's bee mites preferred European honeybees. *Am. Bee J.* 135: 415–420.

Delfinado-Baker, M., Baker, E.W. and Phoon, A.C.G. 1989. Mites (Acari) associated with bees (Apidae) in Asia, with description of a new species. *Am. Bee J.* 129: 609–613.

Delfinado-Baker, M., Underwood, B.A. and Baker, E.W. 1985. The occurrence of *Tropilaelaps* mites in brood nests of *Apis dorsata* and *Apis laboriosa* in Nepal with descriptions of the nymphal stages. *Am. Bee J.* 125: 703–706.

Dinabandhoo, C.L. and Dogra, G.S. 1980. Incidence and control of ectoparasitic mite, *Pyemotes herfsi* Oudemans of the Indian honeybee, *Apis cerana* F. *Am. Bee J.* 120: 44–47.

Donze, G. and Guerin, P.M. 1994. Behavioural attributes and parental care of *Varroa* mites parasitizing honeybee brood. *Behav. Ecol. Sociobiol.* 34: 305–319.

Donze, G. and Guerin, P.M. 1997. Time activity budgets and space structuring by the different life stages of *Varroa jacobsoni* in capped brood of the honeybee, *Apis mellifera. J.Insect Behav.* 10: 371–393.

Dougherty, E.M., Cantwell, G.E. and Kuchinski, M. 1982. Biological control of the greater wax moth (Lepidoptera: Pyralidae) utilizing *in vivo* and *in vitro* propagated baculovirus. *J. Econ. Entomol.* 75: 675–679.

Du, Z.L. and Zhang, Z.B. 1985. Ultra structural change in the hypopharyngeal glands of worker honeybees (*Apis cerana*) infected with sac brood virus. *Zool. Res.* **6**: 155–162.

Dzierzon, J. 1882. *Rational Beekeeping.* Houlston and Sons, London. p. 350.

Eickwort, G.C. 1988. The origins of mites associated with honeybees. In: *Africanized Honeybees and Bee mites.* Needham, G.R., Page, Jr., R.E., Delfinado-Baker, M. and Bowman, C.E. (eds.). Ellis Horwood, Chichester, England. pp. 327–384.

Eickwort, G.C. 1997. Mites: an overview. In: *Honeybee Pests, Predators and Diseases* Morse, R.M.and Flottum, P.K. (eds.). Medina, USA. pp. 241–250.

Ellis, J.D. and Hepburn, H.R. 2006. An ecological digest of the small hive beetle (*Aethina tumida*), a symbiont in honeybee colonies (*Apis mellifera*). *Insects Soc.* **53**: 8–19.

Ellis, J.D. and Munn, P.A. 2005. The worldwide health status of honeybees. *Bee Wld.* **86**: 88–101.

Ellis, J.D., Richards, C.S., Hepburn, H.R.and Elzen, P.J. 2003. Oviposition by small hive beetles elicits hygienic response from Cape honeybees. *Naturwissenschaften.* **90**: 532–535.

Ellis, J.D., Rong, I.H., Hill, M.P., Hepburn, H.R. and Elzen, P. J. 2004. The susceptibility of small hive beetle (*Aethina tumida* Murray) pupae to fungal pathogen. *Am. Bee J.* **144**: 486–488.

Elzen, P. J., Baxter, J.R., Westervelt, D., Randall, C., Cutts, L., Wilson, W.T., Eischen, F.A., Delaplane, K.S. and Hopkins, D. 1999a. Status of small hive beetle in the US. *Bee Cult.* **127**: 28–29.

Elzen, P.J., Baxter, J.R., Westervelt, D., Randall, C., Delaplane, K.S., Cutts, L. and Wilson, W.T. 1999b. Field control and biology studies of a new pest species, *Aethina tumida* Murray (Coleoptera: Nitidulidae), attacking European honeybees in the western hemisphere. *Apidologie.* **30**: 361–366.

Elzen, P.J.,Westervelt, D., Causey, D., Ellis, J., Hepburn, H.R. and Neumann, P. 2002. Method of application of tylosin, an antibiotic for American foul brood control, with effects on small hive beetle (Coleoptera: Nitidulidae) populations. *J. Econ. Entomol.* 95: 1119–1122.

Erickson, E.H., Cohen, A.C. and Cameron, B.E. 1994. Mite excreta: a new diagnostic for varroasis. *Bee Sci.* 3: 76–78.

Evans, J.D. and Hung, A.C. 2000. Molecular phylogenetics and the classification of honeybee viruses. *Archiv. Virol.* 145: 2015–2026.

Eyer, M., Chen, Y.P., Schafer, M.O., Pettis, J. and Neumann, P. 2009. Small hive bettle, *Aethina tumida*, as a potential biological vector of honeybee viruses. *Apidologie.* 40 *(in press)*.

Fakhimzadeh, K. 2000. A rapid field and laboratory method to detect *Varroa jacobsoni* in the honeybee (*Apis mellifera*). *Am. Bee J.* 140: 736–739.

Fakhimzadeh, K. 2001. Acute impact on the honeybee (*Apis mellifera*) after treatment with powdered sugar and $Co_2$ for the control of *Varroa destructor. Am. Bee J.* 141: 817–820.

Feldlaufer, M.F., Kochansky, J.P. and Shimanuki, H. 1998. The use of a sterol inhibitor to control the greater wax moth, *Galleria mellonella. Am. Bee J.* 138: 287–289.

Feldlaufer, M.F., Knox, D.A., Lusby, W.R. and Shimanuki, H. 1993. Antimicrobial activity of fatty acids against *Bacillus larvae*, the causative agent of AFB disease. *Apidologie.* 24: 95–99.

Fiedler, L. 1987. Acephate residues after preblossom treatments: effect on small colonies of honeybees. *Bull. Environ. Contam. Toxicol.* 38: 594–601.

Free, J.B. and Williams, I.H. 1979. Communication by pheromones and other means in *Apis florea* colonies. *J. Apic. Res.* 18: 16–25.

Fries, I. and Camazine, S. 2001. Implications of horizontal and vertical pathogen transmission for honeybee epidemiology. *Apidologie.* 32: 199–214.

Fries, I., Feng, F., Silva, A.D., Slemenda, S.B. and Pieniazek, N.J. 1996. *Nosema ceranae* n. sp. (Microspora, Nosematidae), morphological and molecular characterization of a microsporidian parasite of the Asian honey bee *Apis cerana* (Hymenoptera: Apidae). *European J. Protistol.* 32: 356–365.

Garg, R. and Kashyap, N.P. 1998. Mites and other enemies of honeybees in India. In: *Perspectives in Indian Apiculture.* Mishra, R.C. and Garg, R. (eds.). *Agro Botanica,* New Delhi, India. pp. 264–303.

Garg, R. and Sharma, O.P. 1988. Studies to establish the effective dose of precipitated sulphur against *Tropilaelaps clareae* Delfinado and Baker (Laelapidae: Acarina). *Am. Bee J.* 128: 122–123.

Garg, R., Sharma, O.P. and Dogra, G.S. 1984. Formic acid: an effective acaricide against *Tropilaelaps clareae* Delfinado and Baker (Laelapidae: Acarina) and its effect on the brood and longevity of honeybee. *Am. Bee J.* 124: 736–738.

Gary, N.E., Page, R.E. and Lorenzen, K. 1989. Effects of age of worker honeybees (*Apis mellifera* L.) on tracheal mite (*Acarapis woodi* Rennie) infestation. *J. Exp. Appl. Acarol.* 7: 153–160.

Garzon, S., Charpentier, G. and Kursta, K.E. 1978. Morphogenesis of the Nodamura virus in the larvae of the Lepidoptera, *Galleria mellonella. Arch. Virol.* 56: 61–76.

Genersch, E., Ashiralieva, A. and Fries, I. 2005. Strain and genotype-specific differences in virulence of *Paenibacillus larvae* subsp. *larvae*, a bacterial pathogen causing American foul brood disease in honeybees. *Appl. Environ. Microbiol.* 71: 7551–7555.

Gilliam, M. 1990. Chalk brood disease of honeybee, *Apis mellifera* caused by the fungus, *Ascosphaera apis*: a review of past and current research. In: *Vth Intern. Coll. Inv. Pathol.Microbiol. control.* Adelaide, Australia. pp. 398–402.

Gilliam, M., Taber, S. and Richardson, G.V. 1983. Hygienic behaviour of honeybees in relation to chalk brood disease. *Apidologie.* 14: 29–39.

Gilliam, M., Taber, S. and Rose, J.B. 1978. Chalk brood disease of honeybees, *Apis mellifera* L. : A progress report. *Apidologie.* **9**: 75–89.

Gilliam, M., Lorenz, B.J., Prest, D.B. and Camazine, S. 1993. *Ascosphaera apis* from *Apis cerana* from South Korea. *J. Inv. Pathol.* **61**: 111–112.

Gochnauer, T.A. and Margetts, V.J. 1980. Decontaminating effect of ethylene oxide on honeybee larvae previously killed by chalk brood disease. *J. Apic Res.* **19**: 261–264.

Goebel, R.L. 1984. Bee birds in Queensland. *Australian Beekeeper.* **86**: 137–138.

Goodwin, M. and Haine, H. 1998. Sterilizing beekeeping equipment infected with American foul brood spores. *New Zealand Beekeeper.* **5**: 13.

Govan, V.A., Brozel, V., Allsopp, M.H. and Davison, S. 1998. A PCR detection method for rapid identification of *Melissococcus pluton* in honeybee larvae. *Appl. Environ. Microbiol.* **64**: 1983–1985.

Grant, G.M., Nelson, D.L., Olsen, P.E. and Rice, W.A. 1993. The ELISA detection of tracheal mites in whole honeybee samples. *Am. Bee J.* **133**: 652–655.

Griffiths, D.A. 1988. Functional morphology of the mouth parts of *Varroa jacobsoni* and *Tropilaelaps clareae* as a basis for the interpretation of their life styles. In: *Africanized Honeybees and Bee mites.* Needham, G.R., Page, Jr. R.E., Delfinado-Baker, M. and Bowman, C.E. (eds.). Ellis Horwood, Chichester, England. pp. 479–486.

Grobov, O.F. 1974. Gamasoid mites as parasites of bees. *Veterinarii.* **8**: 78–82.

Gross, O., Tijssen, P., Weinberg, D. and Tal, J. 1990. Expression of densonucleosis virus from GmDNV in *Galleria mellonella* larvae: size analysis and *in vitro* translation of viral transcription products. *J. Inv. Pathol.* **56**: 175–180.

Gruszka, J. 1988. Evaluation of the ether sampling method to detect low level infestations of *Varroa jacobsoni. Beelines.* **5**: 8–14.

Hadisoesilo, S., Otis, G.W. and Meixner, M. 1996. Two distinct populations of cavity nesting honeybees (Hymenoptera: Apidae) in south Sulawesi, Indonesia. *J. Kans. Entomol. Soc.* **68**: 399–407.

Hansen, H. 1984. Methods for determining the presence of the foul brood bacterium, *Bacillus larvae* in honey. *Tidsskrift for Planteavl.* **88**: 325–328.

Hansen, H. and Brødsgaard, C.J. 2003. Control of American foul brood by the shaking method. *Apiacta.* **38**: 140–145.

Hansen, H. and Rasmussen, B. 1990. The sensitiveness of the foul brood bacterium *Bacillus larvae* to heat treatment. In: Proc. Intern. Symp. Recent Res. Bee Pathol. Gent, Belgium. pp. 146–148.

Hanumanthaswamy, B.C. 2000. Natural enemies of honeybees with special reference to Bioecology and Management of greater wax moth, *Galleria mellonella* Linnaeus (Lepidoptera: Pyralidae). Ph.D thesis, University of Agricultural Sciences, Bangalore, India, p. 257.

Hanumanthaswamy, B.C., Rajagopal, D. and Visweswara Gowda, B.L. 2001. Comparative biology of greater wax moth, *Galleria mellonella* on combs of different species of honeybees. *Asian Bee J.* **3**: 27–33.

Haq, M.A., Halliday, R.B., Walter, D.E., Proctor, H.C. and Norton, R.A. 2001. Life cycle and behaviour of coconut mite, *Neocypholaelaps stridulans*. In: *Acarology: Proc. 10th Intern. Congr.* CSIRO Publishing, Melbourne, Australia. pp. 361–365.

Harbo, J.R. and Harris, J.W. 1999. Heritability in honeybees (Hymenoptera: Apidae) of characteristics associated with resistance to *Varroa jacobsoni* (Mesostigmata: Varroidae). *J. Econ. Entomol.* **92**: 261–265.

Harbo, J.R. and Hoopingarner, R. 1997. Honey bees (Hymenoptera: Apidae) in the United States that express resistance to *Varroa jacobsoni* (Mesostigmata: Varroidae). *J. Econ. Entomol.* **90**: 893–898.

Haseman, L. 1948. Further studies with sulfathiazole for control of foul brood. *J. Econ. Entomol.* **41**: 120.

Haseman, L. 1961. How long can spores of American foul brood live? *Am. Bee J.* 101: 298–299.

Haung, Z.H. 1984. Habits of *Galleria mellonella* L. and *Achroia grisella* F. and methods of control. *Insect Knowledge*. 21: 36–39.

Herbert, Jr. E.W., Chitwood, W. and Shimanuki, H. 1986. Effect of a candidate compound on chalk brood disease in New Jersey. *Am. Bee J.* 126: 258–259.

Heyndrickx, M., Vandemeulebroecke, K., Hoste, B., Janssen, P., Kersters, K., de Vois, P., Logan, N.A., Ali, N. and Berkeley, R.C.W. 1996. Reclassification of *Paenibacillus* (formerly *Bacillus*) *larvae* with emended description of *P. larvae* and *P. larvae* subsp. *pulvifaciens*. *Intern. J. Systematic. Bacteriol.* 46: 270–279.

Hirst, S. 1921. On the mites (*Acarapis woodi* Rennie) associated with Isle of Wight bee disease. *Ann. Mag. Nat. Hist.* 7: 509–519.

Hood, W.M. 2004. The small hive beetle, *Aethina tumida:* A review. *Bee Wld.* 85: 51–59.

Hood, M.W. and Miller, G.A. 2003. Trapping small hive beetles (Coleoptera: Nitidulidae) inside colonies of honeybees (Hymenoptera: Apidae). *Am. Bee J.* 143: 405–409.

Hoppe, H. and Ritter, W. 1989. The influence of the Nasanov pheromone on the recognition of house bees and foragers by *Varroa jacobsoni*. *Apidologie*. 19: 165–172.

Hornitzky, M.A.Z. 1986. The use of gamma radiation in the control of honeybee infections. *Australian Beekeeper*. 88: 55–58.

Hornitzky, M.A.Z. and White, B. 2001. Controlling American Foul brood : assessing effectiveness of shaking bees and antibiotic therapy strategies. RIRDC Publication. Kingston, Australia. p. 21.

Hornitzky, M.A.Z.and Willis, P.A. 1983. Gamma radiation inactivation of *Bacillus larvae* to control American foul brood. *J. Apic. Res.* 22: 196–199.

Huang, W.F., Jiang, J.H., Chen, Y.W. and Wang, C.H.2007. A *Nosema ceranae* isolate from the honeybee, *Apis mellifera*. *Apidologie*. 38: 30–37.

Ibay, L.A. 1989. Biology of the two external *Acarapis* species of honeybees, *Acarapis dorsalis* and *Acarapis externus* (Acari: Tarsonemidae). M. S. Thesis. Oregon State University, USA. p. 48.

Ifantidis, M.D. and Rosenkranz, P. 1988. Reproduktion der Bienenmilbe *Varroa jacobsoni* (Acarina: Varroidae). *Entomol. Gener.* 14: 111–122.

Imdorf, A., Bogdanov, S., Ochoa, R.I. and Calderone, N. 1999. Use of essential oils for the control of *Varroa jacobsoni* Oud. in honeybee colonies. *Apidologie.* **30**: 209–228.

Imms, A.D. 1942. On *Braula coeca* Nitsch and its affinities. *Parasitol.* **34**: 88–100.

Jadczak, A.M. 1986. *Honeybee Diseases and Pests.* Maine Dept. of Agric. Food and Rural Resour. USA, p. 24.

Jelinski, M. and Wojtowski, F. 1984. *Melittobia acasta* Walker (Hym: Chalcidoidea, Enlophidae). *Malonany Pasozytczerwia Pszcelego Pr Zegl. Zool.* **28**: 507–511.

Kaneda, T. 1969. Fatty acids in *Bacillus larvae, Bacillus lentimorbus and Bacillus popillae. J. Bacteriol.* **98**: 143–146.

Kanga, L.H.B., Jones, W.A. and James, R.R. 2003. Field trials using the fungal pathogen, *Metarhizium anisopliae* (Deuteromycetes: Hyphomycetes) to control the ectoparasitic mite, *Varroa destructor* (Acari: Varroidae) in honeybee, *Apis mellifera* (Hymenoptera: Apidae) colonies. *J. Econ. Entomol.* **96**: 1091–1099.

Kannan, R. 1996. Prospects for beekeeping at the Palani hills and associated foothill plains with *Apis cerana indica.* In: Proc. Intern. Workshop Reviv. *Apis cerana indica.* Palani hills Conserv. Council, Kodaikanal, Tamil Nadu, India. pp. 15–19.

Kapil, R.P. and Aggarwal, K. 1987. Observations on reproduction and seasonal population trends of *Euvarroa sinhai* (Mesostigmata: Varroidae) in India. In: *Progress in Acarol.*Channa Basavanna, G.P. and Viraktamath, C.A.(eds.). Oxford and IBH Pub. Co. Pvt. Ltd., New Delhi, India. pp. 277–282.

Kapil, R.P. and Sihag, R.C. 1983. Wax moth and its control. *Indian Bee J.* 45: 47–49.

Kapil, R.P., Putatunda, B.N. and Aggarwal, K. 1985. Studies on ectoparasitic mites of *Apis* species. Tech. Rep. Dept. of Zoology, Haryana Agric. University. Hissar, India, p. 191.

Katznelson, H. 1950. The influence of antibiotics and sulfa drugs on *Bacillus larvae*, cause of American foul brood of the honeybee, *in vitro* and *in vivo*. *J. Bacteriol.* 59: 153–155.

Katznelson, H. and Robb, J.A. 1962. The use of gamma radiation from cobalt-60 in the control of diseases of honeybee and the sterilization of honey. *Can. J. Microbiol.* 8: 175–179.

Kastberger, G. and Sharma, D. K. 2000. The predator-prey interaction between blue bearded bee eaters (*Nyctyornis athertoni* Jardine and Selby) and giant honeybees (*Apis dorsata* Fabricius). *Apidologie.* 31: 727–736.

Kitprasert, C.H. 1984. Biology and systematics of the parasitic mite, *Tropilaelaps clareae* Delfinado and Baker (Acarina: Laelapidae). M.Sc. Thesis, Kasetsart University. Nakahom Pathom, Thailand, p. 70.

Knox, D.A., Shimanuki, H. and Caron, D.M. 1976. Ethyl oxide plus oxytetracycline for the control of American foul brood in honeybees. *J. Econ. Entomol.* 69: 606–608.

Koenig, J.P., Boush, G.M. and Erickson, Jr. E.H. 1987. Isolation of the chalk brood pathogen, *Ascosphaera apis* from honeybee (*Apis mellifera*) surfaces, pollen loads and a water source. *Am. Bee J.* 127: 581–583.

Koeniger, N. 1995. Biology of the eastern honeybee, *Apis cerana* (Fabricius 1773). In: *The Asiatic Hive bee: Apiculture, Biology and Role in Sustainable Development in Tropical and Subtropical Asia.* Kevan, P.G. (ed.). Enviroquest Ltd., Canada. pp. 29–30.

Koeniger, N. and Muzaffar, N. 1988. Life span of the parasitic honeybee mite, *Tropilaelaps clareae* on *Apis cerana*, *Apis dorsata* and *Apis mellifera*. *J. Apic. Res.* 27: 207–212.

Koeniger, G., Koeniger, N., Anderson D.L., Lekprayoon, C. and Tingek, S. 2002. Mites from debris and brood cells of *Apis dorsata* colonies in Sabah (Borneo) Malaysia, including new haplotype of *Varroa jacobsoni*. *Apidologie*. **33**. 15–24.

Koeniger, N., Koeniger, G., Gries, M., Tingek, S. and Kelitu, A. 1996. Observations on colony defense of *Apis nuluensis* Tingek, Koeniger and Koeniger 1996 and predatory behaviour of the hornet, *Vespa multimaculata* Perez. *Apidologie*. **27**: 341–352.

Kshirsagar, K.K. and Godbole, S.H. 1974. Indian *Streptococcus pluton*. *Indian Bee J*. **36**: 39.

Kshirsagar, K.K. and Mahindre, D.B. 1975. Some notes on bee predatory wasps in India. *Indian Bee J*. **37**: 4–9.

Kshirsagar, K.K., Chauhan, R.M. and Singh, Y.K. 1974. Detection of *Nosema apis* Zander in *Apis cerana indica* F. in Uttar Pradesh (India). *Indian Bee J*. **36**: 23–24.

Kshirsagar, K.K., Salvi, S.R., Mahindre, D.B. and Chauhan, R.M. 1980. Menthol as an effective acaricidal fumigant. In: Proc. IInd Intern. Conf. Apic. Trop. Clim., New Delhi, India.

Laigo, F.M. and Morse, R.A. 1969. Control of the bee mites, *Varroa jacobsoni* Oudemans and *Tropilaelaps clareae* Delfinado and Baker with Chlorobenzilate. *Philippine Entomologist*. **1**: 144–148.

Landerkin, G.B. and Katznelson, H. 1959. Organisms associated with septicemia in honeybee, *Apis mellifera* L. *Can. J. Microbiol*. **5**: 169–172.

Leephitakrat, S., Wongsiri, S., Lerdthusnee, K. and Aemprapa, S. 1999. Investigation of the efficacy of various strains of *Bacillus thuringiensis* on the lesser wax moth, *Achroia grisella* and greater wax moth, *Galleria mellonella*. *Asian Bee J*. **1**: 1–7.

Lescure, J. 1966. The predatory behaviour of the common toad towards bees. *Ann. Abeille*. **9**: 83–114.

Lindauer, M. 1957. Communication among the honeybees and stingless bees of India. *Bee Wld*. **38**: 3–14 and 34–39.

Lindberg, C.M., Melathopoulos, A.P. and Winston, M.L. 2000. Laboratory evaluation of miticides to control *Varroa jacobsoni* (Acari: Varroidae) a honeybee (Hymenoptera: Apidae) parasite. *J. Econ. Entomol.* **93**: 189–198.

Lindquist, E.E. 1968. An unusual new species of *Tarsonomus* (Acarina: Tarsonemidae) associated with the Indian honeybee. *Can. Ent.* **100**: 1002–1006.

Lindquist, E.E. 1986. The World genera of Tarsonemidae (Acari: Heterostigmata): a morphological, phylogenetic and systematic revision with a re-classification of family group taxa in the Heterostigmata. *Mem. Entomol. Soc. Can.* **136**: 1–517.

Lindstrom, A., Korpela, S. and Fries, I. 2008. Horizontal transmission of *Paenibacillus larvae* spores between honeybee (*Apis mellifera*) colonies through robbing. *Apidologie.* **39**: 515–522.

Liu, T.P. 1985. Scanning electron microscope observations on the pathological changes of Malpighian tubules in the worker honeybee, *Apis mellifera* infected by *Malpighamoeba mellificae. J. Inv. Pathol.* **46**: 125–132.

Liu, T.P. 1995. A possible control of chalk brood and *Nosema* diseases of the honeybee with neem. *Am. Bee J.* **135**: 195–198.

Liu, T.P. 1996. *Varroa* mites as carriers of honeybee chalk brood. *Am. Bee J.* **136**: 655.

Liu, T.P., Mobus, B. and Braybrook, G. 1989. A scanning electron microscope study on the prothoracic tracheae of the honeybee, *Apis mellifera* L. infested by the mite, *Acarapis woodi* (Rennie). *J. Apic. Res.* **28**: 81–84.

Loglio, G. and Penessi, E. 1991. On the use of wheat flour for an ecological control of Varroasis. *Apicolt. Mod.* **82**: 185–192.

Lom, J. 1964. The occurrence of a *Crithidia* species within the gut of the honeybee, *Apis mellifera* L. *Entomophaga. Memoire Hors. Serie.* **2**: 91–93.

Lommell, S.A., Morris, T.J. and Pinnock, D.E. 1985. Characterization of nucleic acids associated with Arkansas bee virus. *Intervirol.* **23**: 199–207.

Maa, T. 1953. An inquiry into the systematics of the tribus Apidini or honeybees (Hym). *Treubia.* **21**: 525–640.

Machado, V.L.L., Gobbi, N. and Simoes, D. 1987. Prey items utilized by *Stelopolybia pallipes* Olivier (Hymenoptera: Vespidae). *Anais da Sociedade Entomologia do Brasil.* **16**: 73–79.

Mackenson, O. 1951. Viability and sex determination in the honeybee (*Apis mellifera* L.). *Genetics.* **36**: 500–509.

Mahadevan, V. 1951. *Acanthaspis siva* Distant: a predator of the Indian honeybee. *Indian Bee J.* **13**: 143–147.

Mahindre, D.B. 1983. Handling rock bee colonies. *Indian Bee J.* **45**: 72–73.

Maistrello, L., Lodesani, M., Costa, C., Leonardi, F., Marani, G., Caldon, M., Mutinelli, F. and Granato, A. 2008. Screening of natural compounds for the control of nosema disease in honeybees (*Apis mellifera*). *Apidologie.* **39**: 436–445.

Manino, A. and Patetta, A. 1987. Wasps. *Apiculture Moderno.* **78**: 245–253.

Martin, S.J. 1994a. Ontogenesis of the mite, *Varroa jacobsoni* Oud. in worker brood of the honeybee, *Apis mellifera* L. under natural conditions. *Exp. Appl. Acarol.* **18**: 87–100.

Martin, S.J. 1994b. Ontogenesis of the mite, *Varroa jacobsoni* Oud. in drone brood of the honeybee, *Apis mellifera* L. under natural conditions. *Exp. Appl. Acarol.* **19**: 199–210.

Martin, S.J. 2004. Acaricide (pyrethroid) resistance in *Varroa destructor.* *Bee Wld.* **85**: 67–69.

Martin, S.J. and Kemp, D. 1997. Average number of reproductive cycles performed by *Varroa jacobsoni* in honeybee (*Apis mellifera*) colonies. *J. Apic. Res.* **36**: 113–123.

Matsuka, M., Watanabe, N. and Takeuchi, K. 1973. Analysis of food of larval drone honeybees. *J. Apic. Res.* 12: 3–7.

Mattila, H.R., Otis, G.W., Daley, J. and Schulz, T. 2000. Trials of Apiguard, a thymol based miticide. Non-target effects on honeybees. *Am. Bee J.* 140: 68–70.

McKee, B.A., Djordjevic, S.P., Goodman, R.D. and Hornitzky, M.A. 2003. The detection of *Melissococcus pluton* in honeybees (*Apis mellifera*) and their products using hemi-nested PCR. *Apidologie.* 34: 19–27.

Meikle, W.G., Mercadier, G., Holst, N., Nansen, C. and Girod, V. 2007. Duration and spread of an entomopathogenic fungus, *Beauveria bassiana* (Deuteromycetes: Hyphomycetes) used to treat *Varroa* mites (Acari: Varroidae) in honeybee (Hymenoptera: Apidae) hives. *J. Econ. Entomol.* 100: 1–10.

Melathopoulos, A.P., Nelson, D. and Clark, K.2004. High velocity electron beam radiation of pollen and comb for the control of *Paenibacillus larvae* subspecies *larvae* and *Ascosphaera apis. Am. Bee J.* 144: 714–720.

Miller, L.K. 1998. Ascoviruses. In: *The Insect Viruses.* Miller, L.K. and Ball, L.A.(eds.). Plenum Press. New York, USA.

Milne, C.P. Jr. 1985. Laboratory tests of honeybee hygienic behaviour and resistance to EFB. *Am. Bee J.* 125: 578–580.

Mishra, R.C., Dogra, G.S. and Gupta, P.R. 1980. Some observations on iridovirus of bees. *Indian Bee J.* 42: 9–10.

Morgenthaler, O.1963. Die Keimung der *Nosema* sporen *Sudwestdeutsche Imkar.* 15: 102–104.

Moritz, R.F.A. 1988. A re-evaluation of the two locus model for hygienic behaviour in honeybees (*Apis mellifera* L.). *J. Heredity.* 79: 257–262.

Morse, R.A. and Flottum, K. 1997. *Honeybee Pets, Predators and Diseases.* Medina, USA. p. 718.

Morse, R.A. and Hooper, T. 1985. *The Illustrated Encyclopedia of Beekeeping.* Blandford Press. Dorset, UK. p. 432.

Mossadegh, M.S. and Birjandi, K.A. 1986. *Euvarroa sinhai* Delfinado and Baker (Acarina: Mesostigmata): a parasitic mite of *Apis florea* F. in Iran. *Am. Bee J.* 126: 684–685.

Mouches, C., Bove, J.M. and Albisetti, J. 1984. Pathogenecity of *Spiroplasma apis* and other spiroplasmas for honeybees in southwestern France. *Ann. Microbiol.* 135: 151–155.

Muerrle, T.M., Neumann, P., Dames, J. F., Hepburn, H.R. and Hill, M.P. 2006. Susceptibility of adult *Aethina tumida* (Coleoptera: Nitidulidae) to entomopathogenic fungi. *J. Econ. Entomol.* 99: 1–6.

Muzaffar, N. and Ahmed, R. 1986. Studies on hornets attacking honeybees in Pakistan. *Pakistan J. Agric. Res.* 7: 59–63.

Nagaraja, N. 1998a. Factors affecting the establishment of European honeybee, *Apis mellifera* L. in Karnataka. Ph.D. Thesis, Bangalore University, Bangalore, India. p. 128.

Nagaraja, N. 1998b. Rate of invasion and infestation of *Tropilaelaps clareae* in sulphur treated and untreated colonies of *Apis mellifera*. *Himalayan. J. Env.Zool.* 12: 131–136.

Nagaraja, N. 2000. Impact of parasitic brood mites, *Tropilaelaps clareae* and *Varroa jacobsoni* on *Apis mellifera* beekeeping in Karnataka. *J. Ecotoxicol. Environ. Monit.* 10: 205–210.

Nagaraja, N. and Rajagopal, D. 1999. Colony establishment, nesting and foraging activity of little honeybee, *Apis florea* F. ( Hymenoptera: Apidae). *J. Ent. Res.* 23: 331–338.

Nagaraja, N. and Rajagopal, D. 2000. Foraging and brood rearing activity of rockbee, *Apis dorsata* F. (Hymenoptera: Apidae). *J. Ent. Res.* 24: 243–248.

Nagaraja, N. and Rajagopal, D. 2001. Biology of brood mite, *Tropilaelaps clareae* in the brood of *Apis mellifera* (Hymenoptera: Apidae). *Asian Bee J.* 3: 20–26.

Nagaraja, N. and Rajagopal, D. 2003a. Effect of neem oil and sugar dust in management of brood mite, *Tropilaelaps clareae* Delfinado and Baker in *Apis mellifera* L. colonies. *Indian Bee J.* 65: 18–23.

Nagaraja, N. and Rajagopal, D. 2003b. Pests, predators and parasites of European honeybee, *Apis mellifera ligustica* Spin. in Karnataka, India. *Indian Bee J.* **65**: 120–127.

Nagaraja, N., Rajagopal, D. and Gavi Gowda, 2003. Resistance mechanism of European honeybee, *Apis mellifera* against brood mite, *Tropilaelaps clareae*. In: Proc. Apimondia XXXVIII Intern. Congr. Ljubljana, Slovenia. pp. 1–7.

Nagaraja, N. and Reddy, C.C. 1996. Occurrence and distribution of phoretic mite, *Neocypholaelaps indica* in the colonies of hive honeybee species. In: Proc. Nat. Symp. Anim. Behav. Karnatak University, Dharwad, India. pp. 22–28.

Nagaraja, N. and Reddy, C.C. 1999. Invasion and infestation of brood parasitic mites, *Varroa jacobsoni and Tropilaelaps clareae* in honeybee colonies. *Indian Bee J.* **61**: 37–41.

Naim, M. and Bisht, D.S. 1972. A simple method of disinfection of honeybee combs against wax moths. *Indian Bee J.* **34**: 70–71.

Nation, J. L., Robinson, F.A., Yu, S.J. and Botton, A.B. 1986. Influence upon honeybees of chronic exposure to very low levels of selected insecticides in their diet. *J. Apic. Res.* **25**: 170–177.

Nelson, D.L. and Gochnauer, T.A. 1982. Field and laboratory studies on chalk brood disease of honeybees. *Am. Bee J.* **122**: 29–34.

Neumann, P. and Elzen, P. J. 2004. The biology of the small hive beetle (*Aethina tumida*) (Coleoptera: Nitidulidae): Gaps in our knowledge of an invasive species. *Apidologie.* **35**: 229–247.

Newton, D.C., Cantwell, G.C. and Bourquin, E.P. 1975. Removal of freeze killed brood as an index of nest cleaning behaviour in honeybee colonies (*Apis mellifera* L.). *Am. Bee J.* **115**: 388–406.

Nyein, M.M. and Zmarlicki, C. 1982. Control of mites in European bees in Burma. *Am. Bee J.* **122**: 638–639.

Okada, I. 1988. Three species of wax moths in Japan. *Honeybee Sci.* **9**: 145–149.

Oldroyd, B.P. 1996. Evaluation of Australian commercial honeybees for hygienic behaviour, a critical character for tolerance to chalk brood. *Australian J. Exp. Agric.* **36**: 625–629.

Oldroyd, B.P. and Wongsiri, S. 2006. *Asian honeybees: Biology, Conservation and Human Interactions.* Harvard University Press, USA. p. 340.

Olsen, P.E., Grant, G.A., Nelson, D.L. and Rice, W. 1990. Detection of American foul brood disease of honeybees using a monoclonal antibody specific to *Bacillus larvae* in an enzyme linked immunosorbent assay. *Can. J. Microbiol.* **36**: 732–735.

Ono, M., Igarashi, T., Ohno, E. and Sasaki, M. 1995. Unusual thermal defense by a honeybee against mass attack by hornets. *Nature.* **377**: 334–336.

Ostiguy, N., Sammataro, D. and Camazine, S. 1999. How to count *Varroa jacobsoni* without going blind: A sane approach. *Am. Bee J.* **139**: 313–314.

Otis, J. and Kralj, J. 2001. Parasitic mites not present in North America. In: *Mites of the Honeybee*, Webster, T.C. and Delaplane, K.S.(eds.). Dadant and Sons, Hamilton, USA. pp. 251–272.

Oudemans, A.C. 1904. Acarological notes; *XIII Entomologische Berchten Vitgegevon Door de Nederlendsche Entomologische Vereeniging.* **1**: 169–174.

Pavlyushin, V.A. 1976. Role of the toxins produced by the fungus, *Beauveria bassiana* (Bals) in the pathogenesis of an experimental mycosis of the bee moth. *Mikologiyai Fitopatologiya.* **10**: 225–227.

Pearce, A.N., Huang, Z.Y. and Breed, M.D. 2001. Juvenile hormone and aggression in honeybees. *J. Insect Physiol.* **47**: 1243–1247.

Peng, Y. and Peng, K. 1979. A study on the possible utilization of immuno-diffussion and immuno-fluorescence techniques as the diagnostic methods for American foul brood of honeybees (*Apis mellifera*). *J. Inv. Pathol.* **33**: 284–289.

Peng, C.Y.S., Fang, Y.Z., Xu, S.Y. and Ge, L.S. 1987. The resistance mechanism of the Asian honeybee, *Apis cerana* F. to an ectoparasitic mite, *Varroa jacobsoni* Oudemans. *J. Inv. Pathol.* **49**: 54–60.

Peng, C.Y.S., Mussen, E., Fong, A., Cheng, P., Wong, G. and Montague, M.A. 1996. Laboratory and field studies on the effects of the antibiotic tylosin on honeybee, *Apis mellifera* L. (Hymenoptera: Apidae): development and prevention of AFB disease. *J. Inv. Pathol.* 67: 65–71.

Pettis, J.S. and Wilson, W.T. 1996. Life history of the honeybee tracheal mite (Acari: Tarsonemidae). *Ann. Enomol. Soc. Am.* 89: 368–374.

Phelan, L.P., Smith, A.W. and Needham, G.R. 1991. Mediation of host selection by cuticular hydrocarbons in the honeybee tracheal mite, *Acarapis woodi* (Rennie). *J. Chem. Ecol.* 17: 463–473.

Phillips, E.F. 1906. Introduction: In: The bacteria of the Apiary with Special Reference to Bee Diseases. White, G.F. (ed.). USDA Tech.Ser.No.14 USA.

Pinnock, D.E. and Featherstone, N.E. 1984. Detection and quantification of *Melissococcus pluton* infection in honeybee colonies by means of enzyme linked immunosorbent assay. *J. Apic. Res.* 23: 168–170.

Pirk, C.W.W., Hepburn, R., Radloff, S.E. and Erlandsson, J. 2002. Defense posture in the dwarf honeybee, *Apis florea. Apidologie.* 33: 289–294.

Prell, H. 1926. Beitrage zur Kenntnis der Amobenseuche der erwachsenen Honigbiene. *Archiv für Bienenkunde.* 7: 113–121.

Qayyum, H.A. and Nabi, A. 1968. Biology of *Apis dorsata. Pakistan J. Sci.* 19: 109–113.

Rajagopal, D. and Kencharaddi, R.N. 2000. Thai sac brood virus disease of Indian honeybee. University of Agricultural Sciences, Bangalore, India. p. 117.

Rajagopal, D. and Nagaraja, N. 1999. Beekeeping status in Karnataka, India. *Asian Bee J.* 1: 50–59.

Rajagopal, D. and Nagaraja, N. 2003. Potentiality of honeybees in increasing crop production. In: *Potentials of Living Resources.* Tripathi, G. and Kumar, A. (eds.). Discovery Pub. House, New Delhi, India. pp. 105–158.

Rajagopal, D., Veeresh, G.K., Chikkadevaiah, Nagaraja, N. and Kencharaddi, R.N. 1999. Potentiality of honeybees in hybrid seed production of sunflower (*Helianthus annuus* L.). *Indian J. Agric. Sci.* **69**: 40–43.

Rahman, A. and Rahman, S. 1995. Efficacy of certain feeding attractants against predatory wasp (*Vespa magnifica* L.). *Plant Health.* **1**: 66–68.

Ramanan, V.R. and Ghai, S. 1984. Observations on the mite, *Neocypholaelaps indica* Evans and its relationship with the honeybee *Apis cerana indica* Fab. and the flowering of Eucalyptus trees. *Entomon.* **9**: 291–292.

Rao, P.V.S., Mohansundaram, M. and Subramanian, T.R. 1972. *Odontomantis micans* Stal (Mantidae: Dictyoptera) as an enemy of *Apis indica* F. the Indian honeybee in Coimbatore. *Indian Bee J.* **34**: 72.

Rath, W., Delfinado-Baker, M. and Drescher, W. 1991. Observations on the mating behaviour, sex ratio, phoresy and dispersal of *Tropilaelaps clareae* (Acari: Laelapidae). *Intern. J. Acarol.* **17**: 201–208.

Ratnieks, F.L.W. 1992. American foul brood: the spread and control of an important disease of the honeybee. *Bee Wld.* **73**: 177–191.

Reddy, M.S. and Reddy, C.C. 1993. Studies on the distribution of nests of giant honeybee (*Apis dorsata* F.). *Indian Bee J.* **55**: 36–39.

Reddy, M.S., Jayaram, G.N. and Nagaraja, N. 2004. Behavioural resistance of the Indian honeybee, *Apis cerana indica* against Thai sac brood virus disease to rejuvenate sustainable beekeeping in Karnataka, India. In: *Bees for New Asia, VII[th] Asian Apic. Assoc. Conf.* Camaya, E.N. and Cervancia, C.R.(eds.). Univ. Philippines, Los Banos, Philippines. pp. 209–212.

Rennie, J., White, P.B. and Harvey, E.J. 1921. Isle of Wight disease in hive bees. *Trans. Roy. Soc. of Edinburg.* **52**: 737–779.

Reuter, G.S. and Spivak, M. 1998. A simple assay for honeybee hygienic behaviour. *Bee Cult.* **126**: 23–25.

Ribbands, C.R. 1953. *The Behaviour and Social Life of Honeybees.* Intern. Bee Res. Assoc. London. p. 352.

Ribière, M., Lallemand, P., Iscache, A.L., Schurr, F., Celle, O., Blanchard, P., Olivier, V. and Faucon, J.P. 2007. Spread of infectious chronic bee paralysis virus by honeybee (*Apis mellifera* L.) faeces. *Appl. Environ. Microbiol.* **73**: 7711–7716.

Rinderer, T.E. and Rothenbuhler, W.C. 1969. Resistance to American foul brood in honeybees. Comparative mortality of queen, worker and drone larvae. *J. Inv. Pathol.* **13**: 81–86.

Ritter, W., Delaitre, N. and Ifantidis, M. 1984. Use of Folbex VA in smoker to control *Varroa* disease. *Apiacta* **19**: 37–39.

Roberts, D.W. and Leger, R.J. 2004. *Metarhizium* spp. cosmopolitan insect pathogenic fungi; mycological aspects. *Adv. App. Microbiol.* **54**: 1–70.

Robinson, G.E. and Page, R.E. 1988. Genetic determination of guarding and undertaking in honeybee colonies. *Nature.* **333**: 356–358.

Rothenbuhler, W.C. 1964. Behavioural genetics of nest cleaning in the honeybees. Responses of F1 and back cross generation to disease killed brood. *Am. Zool.* **4**: 111–123.

Roubik, D.W., Sakagami, S.F. and Kudo, I. 1985. A note on distribution and nesting of the Himalayan honeybee, *Apis laboriosa* Smith (Hymenoptera: Apidae). *J. Kansas Entomol. Soc.* **58**: 746–749.

Ruffinengo, S., Eguaras, M., Floris, I., Faverin, C., Bailac, P. and Ponzi, H. 2005. LD50 and repellent effects of essential oils from Argentinian wild plant species on *Varroa destructor. J. Econ. Entomol.* **98**: 651–655.

Ruttner, F. 1987. *Biogeography and Taxonomy of Honeybees.* Springer-Verlag, Berlin, p. 284.

Ruttner, F., Kauhausen, D. and Koeniger, N. 1989. Position of the red honeybee, *Apis koschevnikovi* (Buttel-Reepen 1906), within the genus *Apis. Apidologie.* **20**: 395–404.

Sammataro, D. 1995. Studies on the control, behaviour and molecular markers of the tracheal mite (*Acarapis woodi* Rennie) of honeybees

(Hymenoptera: Apidae). Ph.D. Dissertation, Ohio State University, Columbus, USA. p. 125.

Samsinak, K., Vobrazkova, E. and Haragsim, O. 1978. *Melittiphis alvearius* Berlese, a little known bee mite. *J. Apic. Res.* **17**: 50–51.

Sandhu, A.S. and Singh, S. 1966. The biology and brood rearing activities of little honeybee, *Apis florea* F. *Indian Bee J.* **22**: 27–35.

Satta, A., Floris, I., Eguaras, M., Cobras, P., Garau, V.L. and Melis, M. 2005. Fomic acid based treatments for control of *Varroa destructor* in Mediterranean area. *J. Econ. Entomol.* **98**: 267–273.

Schirach, A.G. 1771. Historic naturelle de la reine des abeilles, avec l'art de former des essaims. *Le Haye.* **63**: 269.

Schulz, D.J., Sullivan, J.P. and Robinson, G.E. 2002. Juvenile hormone and octopamine in the regulation of division of labour in honeybee colonies. *Horm. Behav.* **42**: 222–231.

Seeley, T.D. 1985. *Honeybee Ecology: A Study of Adaptation to Social Life.* Princeton Univ. Press. USA. p. 202.

Seeley, T.D., Seeley, R.H. and Akratanakul, P. 1982. Colony defense strategies of the honeybees in Thailand. *Ecol. Monographs.* **52**: 43–63.

Sen Sarma, M., Fuchs, S., Werber, C. and Tautz, R. 2002. Worker piping triggers hissing for coordinated colony defense in dwarf honeybee, *Apis florea. Zoologia.* **105**: 215–223.

Shah, F.A. and Shah, T.A. 1988. *Tropilaelaps clareae,* a serious pest of honeybees: flour dusting control for *Varroa* disease in Kashmir. *Am. Bee J.* **128**: 27.

Shah, F.A. and Shah, T.A. 1991. *Vespa velutina,* a serious pest of honeybees in Kashmir. *Bee Wld.* **72**: 161–164.

Sharma, O.P., Garg, R. and Dogra, G.S. 1983. Efficacy of formic acid against *Acarapis woodi* (Rennie). *Indian Bee J.* **45**: 1–2.

Sharma, S.D., Kashyap, N.P., Raj, D. and Sharma, O.P. 1994. Control of the ectoparasitic mite, *Tropilaelaps clareae* Delfinado and Baker infestation with formic acid and sulphur. *Intern. J. Trop. Agric.* **12**: 96–100.

Shimanuki, H. 1990. Bacteria. In: *Honey bee Pests, Predators and Diseases*. Morse, R.A and Nowogrodzki, R. (eds.). Cornell Univ. Press, USA. pp. 27–47.

Shimanuki, H. and Knox, D.A. 1991. Diagnosis of honeybee diseases. *USDA Hand Book*, USA. p. 53.

Shimanuki, H. and Knox, D. 1997. Summary of control methods. In: *Honey Bee Pests, Predators, and Diseases*. Morse, R.A. and Flottum, K. (eds.). Medina, Ohio, USA.

Shimanuki, H., Knox, D.A. and Herbert, E.W. 1970. Fumigation with ethylene oxide to control diseases of honeybees. *J. Econ. Entomol.* **63:** 1062–1063.

Singh, S. 1957. Acarine disease in the Indian bee, *Apis cerana* F. *Indian Bee J.* **19:** 27–28.

Singh, S. 1961. Appearance of American foul brood disease in Indian honeybee (*Apis indica* F.). *Indian Bee J.* **23:** 46–50.

Singh, S. 1962. *Beekeeping in India*. Indian Council of Agricultural Research, New Delhi, India. p. 214.

Skou, J.P. and Holm, S.N. 1980. Occurrence of melanosis and other diseases in the queen honeybee and the risk of their transmission during instrumental insemination. *J. Apic. Res.* **19:** 133–143.

Smirle, M.J., Mark, L.W. and Kenneth, L.W. 1984. Development of a sensitive bioassay for evaluating sub-lethal pesticide effect on the honeybee (*Apis mellifera*) (Hymenoptera: Apidae). *J. Econ. Entomol.* **77:** 63–67.

Smirnov, A.M. 1982. Study of microbial contamination of hives and combs and methods of disinfection. *Apiacta.* **17:** 100–119.

Smith, F.G. 1960. *Beekeeping in the Tropics*. Longman, London. p. 265.

Smith, I.B. Jr. 1978. The bee louse, *Braula coeca* Nitzsch, its distribution and biology on honey bees. M.S thesis, University of Maryland, USA. p. 111.

Smith, A.W., Page, R.E. Jr. and Needham, G.R.1991. Vegetable oil disrupts the dispersal of tracheal mites, *Acarapis woodi* (Rennie) to young host bees. *Am. Bee J.* **131**: 44–46.

Southwick, E.E. and Moritz, R.F.A. 1987. Effects of meteorological factors on defensive behaviour of honeybee. *Intern. J. Meteorol.* **31**: 259–265.

Spangler, H.G. 1984. Attraction of female lesser wax moths (Lepidoptera: Pyralidae) to male produced and artificial sounds. *J. Econ. Entomol.* **77**: 346–349.

Spiltoir, C.F. and Olive, L.S. 1955. A re-classification of the genus, *Pericystis betts. Mycologia.* **47**: 238–244.

Spivak, M. 1996. Honeybee hygienic behaviour and defense against *Varroa jacobsoni. Apidologie.* **27**: 245–260.

Spivak, M. and Downey, D.L. 1998. Field assays for hygienic behaviour in honeybees (Apidae: Hymenoptera). *J. Econ. Entomol.* **91**: 64–70.

Spivak, M. and Gilliam, M. 1993. Facultative expression of hygienic behaviour of honeybees in relation to disease resistance. *J. Apic. Res.* **32**: 147–157.

Spivak, M. and Gilliam, M. 1998. Hygienic behaviour of honeybees and its application for control of brood diseases and *Varroa. Bee Wld.* **79**: 124–134 and 169–186.

Spivak, M. and Reuter, G.S. 1998. Performance of hygienic honeybee colonies in a commercial apiary. *Apidologie.* **29**: 291–302.

Srivastava, S., Kumar, A., Kashyap, N.P., Chandel, Y.S. and Raj, D. 1995. A new device to protect honeybees from the predatory wasps at the hive entrance of *Apis mellifera* colonies. *Uttar Pradesh J. Zool.* **15**: 165–171.

Stejskal, M. 1973. Gregarinida (Protozoa) , Parasiten der honigbiene in den Tropen. *Die Biene.* **109**: 132–134.

Strick, H. and Madel, G. 1988. Transmission of the pathogenic bacterium *Hafnia alvei* to honeybees by the ectoparasitic mite,

*Varroa jacobsoni*. In: *Africanized Honeybees and Bee mites*. Needham, G.R., Page, Jr. R.E., Delifinado-Baker, M. and Bowman, C.E.(eds.). Ellis Horwood, Chichester, England. pp. 462–466.

Sturtevant, A P. and Revell, I.L. 1953. Reduction of *Bacillus larvae* spores in liquid food of honeybees by action of the honey stopper and its relation to the development of American foul brood. *J. Econ. Entomol.* **46**: 855–860.

Subbaiah, M.S. and Mahadevan, V. 1958. *Vespa cincta* Fab: a predator of the hive bees and its control. *Indian Vet. Sci.* **24**: 153–154.

Sumangala, K. and Haq, M. 2002. Eco-biology of *Pseudacarapis indoapis* Lindquist (Acari: Tarsonemidae) : Ontogeny and breeding behaviour. *J. Ent. Res.* **26**: 83–88.

Suryanarayana, M.C. and Subba Rao, K. 1998. *Apis cerana* F. for Indian apicultute and its management technology. In: *Perspectives in Indian Apiculture*. Mishra, R.C. and Garg, R. (eds.). Agro Botonica, New Delhi, India. pp. 66–118.

Svoboda, J. 1962. Arsenic poisoning of bees by industrial fumes. *Sborn. Akad. Zemed Ved. Rostlinna Vyorba.* 1499–1506.

Taber, S. 1982. Bee behaviour: determining resistance to brood diseases. *Am. Bee J.* **122**: 422–423.

Tanada, Y. and Kaya, H.K. 1993. *Insect Pathology*. Academic Press, San Diego, USA. p. 666.

Tanaka, H., Roubik, D.W. Kato, M. Liew, F. and Gunsalam, G. 2001. Phylogenetic position of *Apis nuluensis* of northern Borneo and phylogeography of *Apis cerana* as inferred from mitochondrial DNA sequences. *Insects Soc.* **84**: 44–51.

Tanaka, H., Watanabe, T., Tawara, Hanaki, K., Huchiyama, K., Tominaga, M. and Inaji, R. 1984. An experiment to protect honeybees from chalk brood disease. *Honeybee Sci.* **5**: 117–120.

Tasei, J.N., Carre, S., Bosio, P.G., Debrey, P. and Hariot, J. 1987. Effects of the pyrethroid insecticide, WL 85871 and phosalone on adults and progeny of the leaf cutting bee, *Megachile rotundata* F. pollinators of Lucerne. *Pesticide Sci.* **21**: 119–128.

Tchuenguem, F. 2002. L' activite de butinage des Apoides sauvage sur les fleur de mais a Yaounde et reflexion sur la pollinisotion des graminees tropicales. *Biotechnologie Agronomic Societe et Environ.* **6**: 87–98.

Thapa, R. and Wongsiri, S. 2003. Flying predators of the giant honeybees, *Apis dorsata* and *Apis laboriosa* in Nepal. *Am. Bee J.* **143**: 540–542.

Thomas, R.T.S. and Thomas, A.M.J.S. 1972. Some observations on the behaviour of females of *Philanthus triangulum* (Hymenoptera: Sphecidae). *Tijdschrift Voor Entomologie.* **155**: 123–139.

Thompson, V.C. 1964. Behavioural genetics of nest cleaning in honeybees. Effect of age of bees of a resistant line on their response to disease killed brood. *J. Apic. Res.* **3**: 25–30.

Thompson, V.C. and Rothenbuhler, W.C. 1957. Resistance to American foul brood in honeybees. Differential protection of larvae by adults of different genetic lines. *J. Econ. Entomol.* **50**: 731–737.

Thorstensen, K. 1976. Kalkyngel en soppsykdom hos bier. *Birokteren.* **92**: 14–17.

Tingek, S., Koeniger, G. and Koeniger, N. 1996. Description of a new cavity dwelling species of *Apis* (*Apis nuluensis*) from Sabah, Borneo with notes on its occurrence and reproductive biology (Hymenoptera: Apoidea: Apini). *Sencken Bergiana Biol.* **76**: 115–119.

Tingek, S., Mardan, M., Rinderer, G., Koeniger, N. and Koeniger, G. 1988. Re-discovery of *Apis vetchi* (Maa, 1953), the Sabah honeybee. *Apidologie.* **19**: 97–102.

Topley, E., Davison, S., Leat, N. and Benjeddou, M. 2005. Detection of three honeybee viruses simultaneously by single multiplex reverse transcriptase PCR. *African J. Biotechnol.* **4**: 763–767.

Trump, R.F., Thompson, V.C. and Rothenbuhler, W.C. 1967. Behavioural genetics of nest cleaning in honeybees. Effect of previous

experience and composition of mixed colonies on response to disease killed brood. *J. Apic. Res.* 6: 127–131.

Underwood, B.A. 1986. The natural history of *Apis laboriosa* in Nepal. M.S. thesis, Cornell University. Ithaca, USA.

Valdes, T. 1974. The pathogenecity of the fungus, *Metarhizium anisopliae* in larvae of *Galleria mellonella. CIARCC.* 4: 5–6.

Van Lawick-Goodall, J. 1968. The behaviour of free living chimpanzees in the Gombe stream reserve. *Anim. Behav. Mon.* 1: 159–311.

Varma, S.K. and Joshi, N.K. 1988. Immunity of *Apis cerana indica* to the Thai sac brood virus. *Indian Bee J.* 50: 39–40.

Verma, L.R., Rana, B.S. and Savithri, V. 1986. Breeding for resistance against Sac brood virus disease of *Apis cerana.* Indian Council of Agricultural Research, New Delhi, India.

Viraktamath, S. 1989. Incidence of greater wax moth, *Galleria mellonella* L. in three species of honeybees. *Indian Bee J.* 51: 139–140.

Viraktamath, S., Basalingappa, S. and Lingappa, S. 2005. Biology and seasonal incidence of the braconid wasp, *Apanteles galleriae* and its parasitization on greater wax moth, *Galleria mellonella. Indian Bee J.* 67: 182–187.

Wager, B.R. and Breed, M.D. 2000. Does honeybee sting alarm pheromone give orientation information to defensive bees? *Ann. Entomol. Soc. Am.* 93: 1329–1332.

Wallace, F.G. 1966. The trypanosomatid parasites of insects and arachnids. *Exp. Parasitol.* 18: 124–193.

White, G.F. 1917. Sac brood. *USDA. Bull.* 431. p. 55.

White, G.F. 1920. American foul brood. *USDA. Bull.* 809. p. 54.

Williams, J.R. Peng, C.Y.S. Chuang, R.Y., Doi, R.H. and Mussen, E.C. 1998. The inhibitory effect of azadirachtin on *Bacillus subtilis, Escherichia coli* and *Paenibacillus larvae,* the causative agent of American foul brood in the honeybee, *Apis mellifera* L. *J. Inv. Pathol.* 72: 252–257.

Williams, G.R., Sampson, M.A., Shutler, D. and Rogers, R.E.L. 2008. Does fumagillin control the recently detected invasive parasite, Nosema ceranae in western honeybees (*Apis mellifera*). *J. Inv. Pathol.* **99**: 342–344.

Wongsiri, S., Lekprayoon, C., Thapa, R., Thirakupt, K., Rinderer, T.E., Sylvester, H.A., Oldroyd, B.P.and Boocham, U.1997. Comparative biology of *Apis andreniformis* and *Apis florea* in Thailand. *Bee Wld.* **78**: 23–35.

Woodraw, A.W. and Holst, E.C. 1942. The mechanism of colony resistance to American foul brood. *J. Econ. Entomol.* **35**: 327–330.

Woyke, J. 1984a. Increase in life span, unit honey productivity and honey surplus with fumagillin treatment of honeybees. *J. Apic. Res.* **23**: 209–212.

Woyke, J. 1984b. Survival and prophylactic control of *Tropilaelaps clareae* infesting *Apis mellifera* colonies in Afghanistan. *Apidologie.* **15**: 421–434.

Woyke, J. 1985. Further investigation into control of parasitic bee mite *Tropilaelaps clareae* without medication. *J. Apic. Res.* **24**: 250–254.

Woyke, J. 1987a. Infestation of honeybee (*Apis mellifera*) colonies by the parasitic mites *Varroa jacobsoni* and *Tropilaelaps clareae* in South Vietnam and results of chemical treatment. *J. Apic. Res.* **26**: 64–67.

Woyke, J. 1987b. Length of stay of parasitic mite *Tropilaelaps clareae* outside sealed honeybee brood cells as basis for its proper control. *J. Apic. Res.* **26**: 104–109.

Woyke, J. 1987c. Length of successive stages in development of the mite, *Tropilaelaps clareae* in relation to honeybee brood age. *J. Apic. Res.* **26**: 110–114.

Woyke, J.1987d. Comparative population dynamics of *Tropilaelaps clareae* and *Varroa jacobsoni* mites on honeybees. *J. Apic. Res.* **26**: 196–202.

Woyke, J. 1989. Change in shape of *Tropilaelaps clareae* females and the onset of egg laying. *J. Apic. Res.* **28**: 196–200.

Woyke, J. 1994. Mating behaviour of the parasitic honeybee mite *Tropilaelaps clareae*. *Exp. Appl. Acarol.* **18**: 723–733.

Woyke, J. 1996. Adult *Tropilaelaps* male can feed and survive for two weeks. In: Proc. IIIrd Asian Apic. Assoc Conf. Bee Res.Beekeep. Dev. Hanoi, Vietnam. pp. 198–201.

Woyke, J., Wilde, J. and Reddy, C.C. 2004. Open-air-nesting honey bees *Apis dorsata* and *Apis laboriosa* differ from the cavity-nesting *Apis mellifera* and *Apis cerana* in brood hygiene behaviour. *J. Inv. Pathol.* **86**: 1–6.

Woyke, J., Wilde, J., Reddy, C.C. and Nagaraja, N. 2005. Periodic mass flights of giant honeybee, *Apis dorsata* performed in successive days at two nesting sites in different environmental conditions. *J. Apic. Res.* **44**: 180–189.

Woyke, J., Wilde, J., Wilde, M., Sivaram, V., Cervancia, C., Nagaraja, N., and Reddy, M.S. 2008. Comparison of defense body movements of *Apis laboriosa*, *Apis dorsata dorsata* and *Apis dorsata breviligula* honey bees. *J. Insect Behav.* **21**: 481–494.

Wu, Y. and Kuang, B. 1987. Two species of small honeybee: a study of the genus *Micrapis*. *Bee Wld.* **68**: 153–155.

Yusof, M.R. and Ibrahim, R. 1995. Current status of pests and diseases of the honeybee, *Apis cerana* F. in peninsular Malaysia. In: *The Asiatic Hive bee: Apiculture, Biology and Role in Sustainable Development in Tropical and Subtropical Asia*. Kevan, P.G.(ed.). Enviroquest Ltd., Canada. pp. 185–190.

Zander, E. 1909. Tierische parasiten als Krankheitserreger bei der Biene. *Leipz. Bienenz. Jahrg.* **10**: 147–150, 11: 164–166.

# INDEX

Printed in Dunstable, United Kingdom

85055787R00127